Computational Methods for Electromagnetics and Microwaves

WILEY SERIES IN MICROWAVE AND OPTICAL ENGINEERING

KAI CHANG, Editor
Texas A & M University

INTRODUCTION TO ELECTROMAGNETIC COMPATIBILITY
Clayton R. Paul

OPTICAL COMPUTING: AN INTRODUCTION
Mohammad A. Karim and Abdul Abad S. Awwal

COMPUTATIONAL METHODS FOR ELECTROMAGNETICS AND MICROWAVES
Richard C. Booton, Jr.

FIBER-OPTIC COMMUNICATION SYSTEMS
Govind P. Agrawal

Computational Methods for Electromagnetics and Microwaves

RICHARD C. BOOTON, JR.
Department of Electrical and Computer Engineering
The University of Colorado at Boulder
Boulder, Colorado

A WILEY-INTERSCIENCE PUBLICATION
JOHN WILEY & SONS, INC.
NEW YORK / CHICHESTER / BRISBANE / TORONTO / SINGAPORE

Library of Congress Cataloging in Publication Data:

Booton, Richard C., 1926–
Computational methods for electromagnetics and microwaves/Richard C. Booton, Jr.
p. cm. —(Wiley series in microwave and optical engineering)
"A Wiley-Interscience publication."
Includes bibliographical references.
ISBN 0-471-52804-8
1. Electromagnetism—Data Processing. 2. Microwaves—Data processing. I. Title. II. Series.

QC760.54B66 1992 91-34450
621.3—dc20 CIP

Printed in the United States of America

10 9 8 7 6 5 4 3 2 1

Printed and bound by Quinn - Woodbine, Inc.

Contents

Preface

This book is based on my experience teaching a beginning graduate course in the Electrical Engineering Department at the University of California, Los Angeles, and in the Electrical and Computer Engineering Department at the University of Colorado at Boulder and is an expansion of notes prepared for this course. This book is intended to serve as a textbook, not as a reference. My experience is that there are several excellent books that can be used as references and assigned as supplemental reading, but that a real need exists for a textbook that can be used for an electrical engineering course. I have deliberately kept the length and depth of the coverage to what can be covered in a one-semester course.

I firmly believe that numerical methods can be learned effectively only by solving realistic problems, and such solutions require that the student write and run computer programs. Although useful programs can be written in BASIC, PASCAL, and other languages, almost all electrical engineers use FORTRAN, and examples in this book use Microsoft FORTRAN. Although the question of what computer language ought to be taught and used is a controversial subject with many people, it seems to me not to matter much what language is used. Far more important is the understanding by the student of the algorithms and how they relate to the physical problems. In a single university quarter or semester, three or four good computer projects can be carried out. I have included what I think are relevant project descriptions in the appropriate chapters.

The enterprising instructor can easily create others. The essential point is that the student must write and use a workable program. Although later in his career he probably will use professionally written "canned" programs, the emphasis here is on his understanding the basic algorithms.

The emphasis here is on electromagnetic and microwave problems and the fundamental algorithms that can be used as the basis for computer programs that produce useful numerical results. Many such programs can be run on personal computers. Of course, more memory and speed are needed for complex problems. I spend little time with students on software niceties, and the better students seem to pick up such capabilities elsewhere.

I wish to express my deep appreciation to four people: Zdeněk Kopal, who first taught me the fundamentals of numerical methods many years ago at MIT; David Chang, who reawakened my interest in numerical methods for electromagnetics; Nicolaos Alexopoulos, who encouraged me to create and teach the course at UCLA; and especially to my wife, Patricia, both for her general encouragement and for her toleration of my many hours at the computer in our bedroom. Also, I wish to thank Kai Chang for his decision to include this book in his series and for his helpful suggestions on material to be included.

RICHARD C. BOOTON, JR.

CHAPTER ONE

Introduction to Numerical Methods

1.1. ELECTROMAGNETIC PROBLEMS CONSIDERED IN THIS BOOK

The numerical solution of partial differential equations and of integral equations has a long history and a variety of methods have been developed. Electromagnetics leads to such equations, and numerical studies in electromagnetics have been responsible for much of the current development of improved and more powerful methods. The three fundamental methods of finite differences, finite elements, and moment methods are concentrated on here, with brief coverage of some other techniques. Emphasis is on the electromagnetic problems and the algorithms that can be used as a basis for computer programs that yield useful numerical results. The programs described in this book are written in Microsoft FORTRAN and have been run on several personal computers using the Microsoft MS-DOS operating system. Other languages, such as BASIC and PASCAL, can be used, although special modifications will be required when complex variables are involved. With appropriate modifications if different FORTRAN versions are used, the programs can be run on almost any personal computer, workstation, or other computer. Adequate memory and processor speed are required to achieve high accuracy. Real physical problems involve three spatial variables and one time variable. Because the full treatment of such problems uses large memory and long run times, wherever possible simplification to one or two spatial variables is carried out.

The principal engineering applications of numerical methods to electromagnetics are to guided waves, antennas, and scattering. Guided waves probably offer the easiest introduction to most of the numerical methods. The analysis of TEM waves on two-conductor transmission lines leads to static calculations of capacitance, which are used to illustrate the three basic methods. Analysis of waveguides leads to eigenvalue problems which we will solve with both a classical method and a newer method. The first uses algebraic eigenvalue techniques coupled with the finite-difference method, and the second approach uses the time-domain finite-difference method coupled with the discrete Fourier transform. Except for the latter method, these methods are classical methods with long histories.

The analysis of microstrip and similar transmission lines, which are becoming increasingly important in microwave circuits, is a more difficult problem. Because microstrip and similar lines involve nonuniform dielectrics, they cannot support a TEM wave. At low frequencies, the dominant mode can be approximated by a TEM wave. This approximation leads to a method of analysis referred to as quasi-TEM or quasi-static, which is essentially the same as the static analysis of two-conductor lines with a uniform dielectric. At higher frequencies, this approximation breaks down and more sophisticated methods are needed. Accurate solution of such problems results from the combination of the moment method with spectral (Fourier series and Fourier transform) techniques. In the last few years, a number of other methods have been used for such problems.

Another major class of electromagnetic problems involves analysis of scattering of waves. The simplest scattering problem involves a plane wave in free space striking a body, frequently taken as a perfect conductor. The incident wave causes currents in the body and these currents create a second wave called the scattered wave. We shall illustrate scattering calculations for a problem of this type. The analysis of the effects of an obstacle in a waveguide offers another example of scattering.

1.2. BASIC NUMERICAL METHODS

The reader is assumed to have a basic background in numerical methods, such as is provided in the undergraduate course at many universities and covered by a number of excellent texts [1]–[3]. The most important topic in this background is the solution of systems of linear algebraic equations. More advanced topics can be found in several numerical texts [4]–[5].

Electromagnetic problems are sometimes described by differential equations, sometimes by integral integrations, and sometimes by minimization of an integral such as the energy integral. The unknown function is usually continuous and depends on continuous independent variables. Computers clearly can handle only a finite set of numbers, whereas the differential and integral equations that describe electromagnetic fields, or indeed almost any physical variable, involve an infinite set of numbers. By one means or another, the more accurate continuous equations must be manipulated to produce equations that involve a finite set of numbers. The final equations usually are algebraic and in electromagnetics, usually linear, and thus solution of a set of linear algebraic equations is involved. Although in this book we consider only situations with passive components, which lead to linear algebraic equations, if active devices such as transistors are involved, nonlinear equations result. Solution methods must then be modified, almost always through use of iterative methods.

Use of any of the numerical methods considered here involves three steps:

1. Preprocessing, to derive the coefficients in the algebraic equations
2. Solution of the algebraic equations
3. Interpretation of the results

The various numerical methods involve different amounts of theoretical effort and different computational efficiencies. As we shall see, increasing the efficiency of the computational process usually

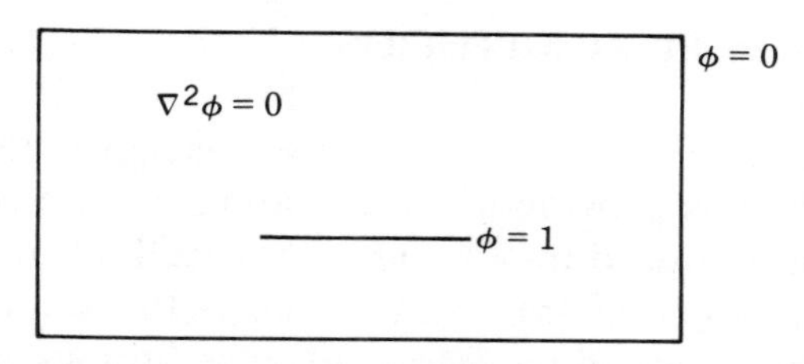

FIGURE 1-1. Two-dimensional potential problem.

must be paid for with more preprocessing effort and more use of electromagnetic theory.

Although, as indicated earlier, a variety of such methods have been developed, three seem to be the most fundamental: the finite-difference method, the finite-element method, and the moment method. As an example, consider the two-dimensional potential problem illustrated in Fig. 1-1. The potential $\phi(x, y)$ that satisfies Laplace's equation and certain boundary conditions on the two conductors is to be determined numerically. The *finite-difference method* concentrates on a finite mesh of points arranged in a rectangular grid or mesh, as shown in Fig. 1-2. By techniques to be discussed in Chapter 2, Laplace's equation is used to establish a set of linear algebraic equations whose solution gives approximate values for the potential on the mesh points. The closely related *finite-element method* divides the region of interest into a number of subregions (the finite elements that give the method its name), as illustrated in Fig. 1-3. The field in each element is approximated by a simple algebraic expression, usually a low-degree polynomial, and by techniques to be discussed later, the values of the field on a

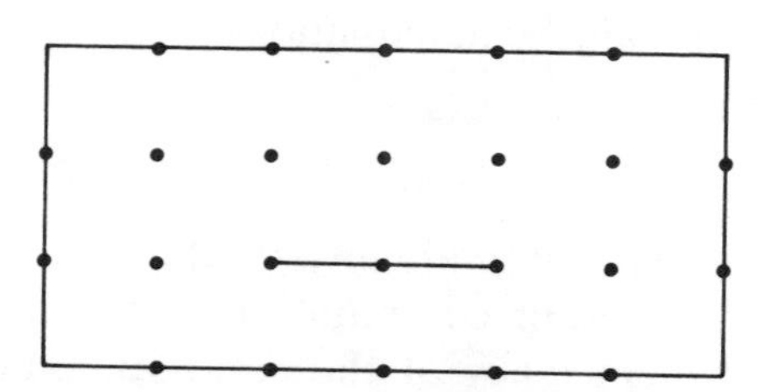

FIGURE 1-2. Finite-difference mesh.

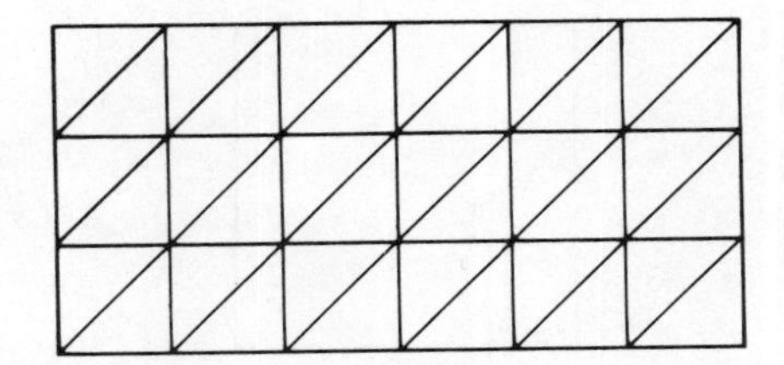

FIGURE 1-3. Triangular finite elements.

finite set of nodal points are found as the solution of a set of linear algebraic equations.

The third method is the *moment method*, sometimes referred to as the method of moments. This method involves more sophisticated mathematics and in return achieves greater computational efficiency. For the potential problem illustrated here, the potential is expressed as an integral over the center conductor of an integrand, which is the product of the charge density and a function known as the *Green's function*. Much of the effort in using the moment method involves calculation of the Green's function, which frequently is determined as the sum of an infinite series. The charge density is expressed as a linear combination of a set of known functions (known as *basis* or *expansion functions*). By methods discussed in the following section, a set of linear algebraic equations are determined whose solution is the set of coefficients in the expansion.

Many other methods, some of which are variations on these basic methods, have been developed. Although we devote one chapter to consideration of one additional method, even a much larger book would be hard-pressed to cover all the methods that have been utilized. More detail on the methods covered here and on other methods can be found in several books [6]–[11].

With any numerical method, maximum use should always be taken of symmetry to reduce memory requirements and computing time. For example, the two-dimensional potential problem illustrated in Fig. 1-1 has one axis of symmetry. Therefore, for finite-difference and finite-element methods, one need consider only one-half of the region, as illustrated in Fig. 1-4. At the artificial boundary caused by the cut along the axis of symmetry, appropriate

FIGURE 1-4. Use of symmetry for potential problem.

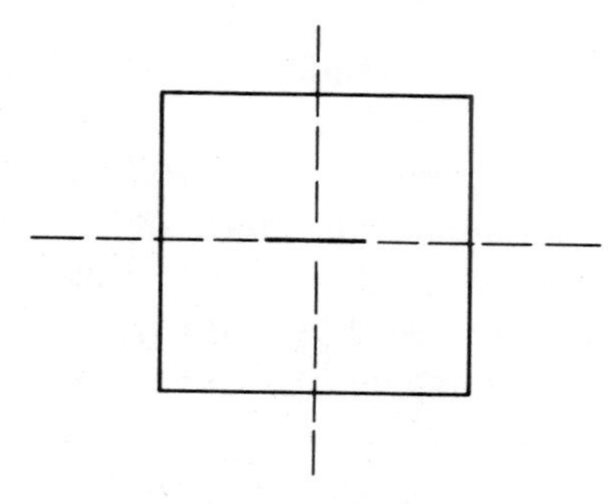

FIGURE 1-5. Problem with two degrees of symmetry.

boundary conditions must be imposed. Moment and other methods also can take advantage of the symmetry, by means to be discussed later. Higher degrees of symmetry may exist. For example, Fig. 1-5 illustrates a situation with two axes of symmetry. If the region is cut along each axis of symmetry, only one-fourth of the region need be considered. Although rare, sometimes even more symmetry exists. In Chapter 2, a situation involving three axes of symmetry will be analyzed.

1.3. SOLUTION OF ALGEBRAIC EQUATIONS

The systems of linear equations to be encountered in the solution of electromagnetic problems fall into two classes. In the first, most of the possible terms are zero and the associated matrix is described as sparse. Such systems frequently are solved by iterative methods such as Gauss–Seidel [12], but powerful methods for direct solution exist [13]. As we shall see when considering discretizing methods such as finite differences, frequently the matrix terms are not initially computed and stored but are computed when required. This is an advantage when dealing with very large matrices.

Some methods, in particular the moment method, involve matrices that in general have no nonzero terms and are referred to as *dense matrices*. Although indirect methods are sometimes used with dense matrices, direct methods are more commonly used. Many algorithms exist for the direct solution of such equations, many of which are variations of Gaussian elimination. All useful algorithms utilize pivoting, which involves interchange of rows or columns to avoid division by zero or by a small number. *Partial pivoting* as usually implemented utilizes interchange of rows, and *full pivoting* interchanges rows and columns. Partial pivoting algorithms are usually adequate and are commonly used. Sometimes the matrix is nearly singular, referred to as *ill-conditioned*, and full-pivoting algorithms are preferred. Most computing networks permit access to excellent matrix inversion routines, and for stand-alone computers without such routines available, excellent routines have been described [14].

1.4. ACCURACY CONSIDERATIONS AND RICHARDSON EXTRAPOLATION

As discussed earlier, we typically convert the original differential or integral equation into a finite set of algebraic equations either by concentrating on a finite set of nodal values or by expanding the unknown function in a finite sum. At least until we have to worry about round-off error, we expect the results we calculate to become more accurate as the number N of unknowns increases. Convergence may be slow, however, and the difference between one value of N and the next is usually small. An approach due to Richardson frequently is useful. This extrapolation method is described in many numerical texts in connection with what is called *Romberg* integration but is a very broadly applicable technique [15].

One approach to using *Richardson extrapolation* is to successively halve the separation between points or double the number of functions. It is reasonable to assume that results calculated with one level of quantization are at least as accurate as the extent to which the results calculated agree with the results at the previous level of quantization. We would like to continue in this manner, successively doubling the number of points or functions until we achieve the accuracy we desire. However, memory requirements and computing time increase rapidly as the number of points or functions increase, and we may not be able to achieve the accuracy we desire in this manner except for simple problems. Consider the results of a series of finite-difference calculations in the following table.

h	Calculation
1	70.83350
0.5	59.87117
0.25	55.67812
0.125	53.79773

Clearly, we could continue to decrease h because the results of the calculation have not converged to an accurate enough result. Notice, however that there is a clear trend in the results, which is made even clearer in Fig. 1-6. Extrapolation to $h = 0$ is suggested

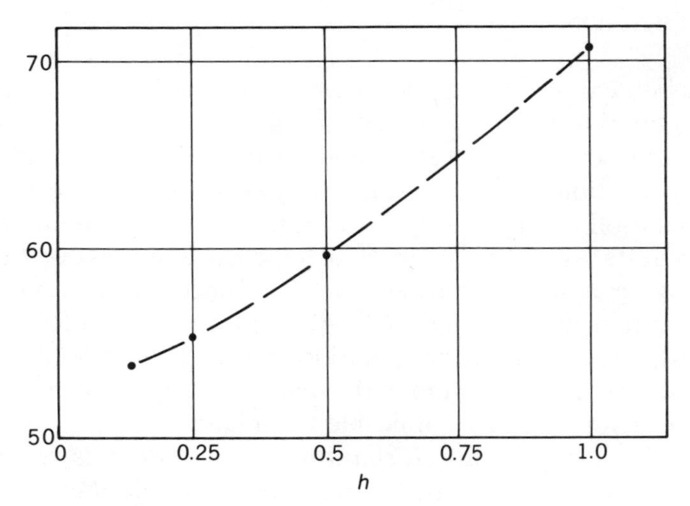

FIGURE 1-6. Graph of calculated results.

by the figure. A numerical extrapolation developed by Richardson based upon a polynomial fit to the calculated numbers will now be explained. Richardson extrapolation has a wide range of application in numerical analysis.

As a first approximation, it is reasonable to assume that the error for the smallest values of h is proportional to h, which means that the calculated values satisfy

$$C(h) = a + bh. \tag{1-1}$$

If calculations are carried out for the two smallest values of h, s, and $2s$, then

$$\begin{aligned} a + bs &= C(s) \\ a + 2bs &= C(2s), \end{aligned} \tag{1-2}$$

from which, by simple algebra, we can compute

$$a = 2C(s) - C(2s). \tag{1-3}$$

The value of a is, of course, the value of the polynomial for h equal to zero, and hence the desired result. This extrapolation uses the last two computed values. Since we have more than two computed values, we should be able to do better by using a higher-degree polynomial that agrees with the computed values. In the calculation above we used two computed values and a first-degree polynomial. In general, the degree of the polynomial that we can use is one less than the number of computed values that we utilize.

For three and more points, formulas similar to (1-3) can be derived. One fits a polynomial to the computed results, calculates the coefficients in that polynomial, and then evaluates the polynomial at zero. A more systematic approach can be used, however. Suppose that the calculated result is expressed as a power series in h in the form

$$C(h) = \sum_{n=0}^{\infty} a_n h^n. \tag{1-4}$$

A first extrapolation can be defined by the formula

$$E_1(h) = 2C(h) - C(2h). \tag{1-5}$$

Substitution of this expression into (1-4) gives

$$E_1(h) = \sum_{n=0}^{\infty} a_n\left[2h^n - (2h)^n\right], \tag{1-6}$$

which, because the coefficient of the first power of h is zero, can be written

$$E_1(h) = a_0 + \sum_{n=2}^{\infty} a_n\left[2h^n - (2h)^n\right]. \tag{1-7}$$

Note that the leading term in the error is a multiple of h^2. If we apply a similar process to extrapolation of E_1, the result is a second extrapolation defined by

$$E_2(h) = \frac{4E_1(h) - E_1(2h)}{3}. \tag{1-8}$$

If we substitute the series into this expression, we will find that the coefficient of the second power of h is now zero. We can continue this process and define the general extrapolation as

$$E_n(h) = \frac{2^n E_{n-1}(h) - E_{n-1}(2h)}{2^n - 1}. \tag{1-9}$$

The number of levels of extrapolation is clearly limited by the number of calculated values we have. Applications of this process to the values in the preceding table leads to the results shown in the following table.

h	Calculation	Extrapolations		
		E_1	E_2	E_3
1	70.83350			
0.5	59.87117	48.90884		
0.25	55.67812	51.48507	52.34381	
0.125	53.79773	51.91734	52.06143	52.02109

One would probably conclude from these results that the true value of the solution is approximately 52.0 with a probable error of 0.1. More extensive calculations on this problem indicate that the solution to three significant figures is indeed 52.0.

This extrapolation process can be inserted into the computer program that computes the original results. One forms a matrix, say $Q(i, j)$, and inserts the calculated results in the first column. The other extrapolated results can then be calculated by a simple algorithm, which in FORTRAN can be written as follows:

```
do i=1,4
      C=1
      do j=2,i
            C=2*C
            Q(i,j)=(C*Q(i,j-1)-Q(i-1,j-1))/(C-1)
      end do
end do
```

With any set of calculated results the most accurate extrapolated values are the diagonal values $Q(j, j)$ and for the example above, the value to use is $Q(4, 4)$. The approach described above in effect passes a polynomial through the calculated results and evaluates this polynomial at $h = 0$. The same result could be obtained by explicitly deriving the polynomial through use of Lagrange interpolation formulas.

In some special applications, such as integration with the trapezoidal method the error series can be shown to contain only even powers of h and the algorithm should be suitably modified to take advantage of this knowledge. The resulting method is known as *Romberg integration*. Usually, however, we have no such knowledge about the error and all powers of h should be included.

1.5. INTEGRATION EXAMPLE

This section considers a simple example that illustrates several of the points discussed thus far. The potential ϕ caused by the charge density ρ is

$$\phi(x, y, z) = \iiint \frac{\rho(x', y', z')}{4\pi\epsilon R} \, dx' \, dy' \, dz', \tag{1-10}$$

where

$$R = \sqrt{(x - x')^2 + (y - y')^2 + (z - z')^2}. \tag{1-11}$$

If the charge is confined to a thin flat plate in the plane $z = 0$ and has a surface charge density of ρ_s, the potential in that plane is

$$\phi(x, y) = \iint \frac{\rho(x', y')}{4\pi\epsilon\sqrt{(x - x')^2 + (y - y')^2}} \, dx' \, dy'. \tag{1-12}$$

Consider now the special case where the plate is a square with sides a and the square is small enough that the charge density can be assumed to be constant over the square. Then

$$\phi(x, y) = \frac{\rho}{4\pi\epsilon} \int_{-a/2}^{a/2} \int_{-a/2}^{a/2} \frac{1}{\sqrt{(x - x')^2 + (y - y')^2}} \, dx' \, dy', \tag{1-13}$$

and the potential at the center of the square is

$$\phi(0, 0) = \frac{\rho}{4\pi\epsilon} \int_{-a/2}^{a/2} \int_{-a/2}^{a/2} \frac{1}{\sqrt{(x')^2 + (y')^2}} \, dx' \, dy'. \tag{1-14}$$

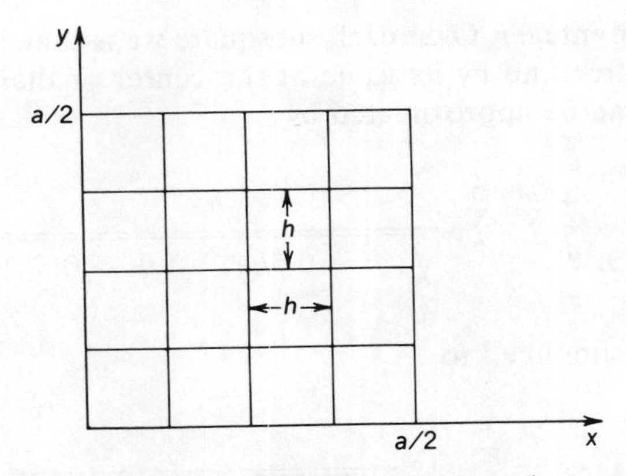

FIGURE 1-7. Division of integration region into squares.

The integral that we now want to evaluate numerically is

$$\mathrm{INT} = \frac{1}{a}\int_{-a/2}^{a/2}\int_{-a/2}^{a/2}\frac{1}{\sqrt{(x')^2+(y')^2}}\,dx'\,dy', \tag{1-15}$$

where we have normalized with respect to a so that INT is a pure number. First let us take advantage of the two degrees of symmetry and evaluate

$$\mathrm{INT} = \frac{4}{a}\int_{0}^{a/2}\int_{0}^{a/2}\frac{1}{\sqrt{(x')^2+(y')^2}}\,dx\,dy'. \tag{1-16}$$

To develop an algorithm for calculation of INT, divide the square into subsquares each with sides of h, as indicated in Fig. 1-7. The largest value of h that we can use is

$$h = \frac{a}{2} \tag{1-17}$$

and in general

$$h = \frac{a}{2M}, \tag{1-18}$$

where M is an integer. Over each subsquare we assume that we can replace the integrand by its value at the center of that subsquare. Then (1-16) can be approximated by

$$\text{INT} = \frac{4}{a} \sum_{i=1}^{M} \sum_{j=1}^{M} \frac{h^2}{\sqrt{(ih - 0.5h)^2 + (jh - 0.5h)^2}}, \tag{1-19}$$

which can be simplified to

$$\text{INT} = \frac{2}{M} \sum_{i=1}^{M} \sum_{j=1}^{M} \frac{1}{\sqrt{(i - 0.5)^2 + (j - 0.5)^2}}. \tag{1-20}$$

To complete the algorithm, we should utilize in succession values for M of $1, 2, 4, \ldots$ up to some value determined by how much accuracy we wish in the final result and how much time we wish to spend in the computation. Also, we should apply Richardson extrapolation to the computed results. A simple computer program to carry out this process follows.

FORTRAN Program for Evaluation of the Potential Integral

```
      program integral
c     evaluation of potential integral
      integer M,i,j,k,N
      real S,Q(10,10)

      print *,'how many values of M?'
      read*,N
      print*
      print*     Potential integral'
      print*,'   Evaluated with pulse functions'
      print*
      print*,'   M    integral   extrapolation'
      print*
      M=1
      do k=1,N
```

```
   S=0
   do j=1,M
      do i=1,M
         S=S+1/(sqrt((i-0.5)*(i-0.5)+
                (j-0.5)*(j-0.5)))/M*2.
         Q(k,1)=S
      end do
   end do
   C=1
   do i=2,k
      C=C*2
      Q(k,i)=(C*Q(k,i-1)-Q(k-1,i-1))/(C-1)
   end do
   print*,M,Q(k,1),Q(k,k)
   M=M*2
end do
end
```

Results of a computer run with eight values of M are shown in the following table.

M	INT	Extrapolations
1	2.828427	2.828427
2	3.150529	3.472631
4	3.330882	3.524102
8	3.426362	3.525561
16	3.475469	3.525497
32	3.500366	3.525495
64	3.512901	3.525492
128	3.519191	3.525496

Note that the direct computations have not converged well even by a value for M of 128, whereas the extrapolated values have clearly converged to a value of 3.52549 by M of 16. The sixth decimal digit exhibits round-off error. This integral can be evaluated analytically and to six significant figures agrees with our numerical evaluation. The integrand in the continuous integral has a singularity at the origin, and the analytical solution must handle this in a limiting fashion. In our numerical solution, the points at which we evaluated the integrand did not include the origin and hence we avoided the problem.

PROBLEMS

1. Use of the Richardson extrapolation method with three computed results for $h = 4s$, $2s$, and s leads to an extrapolation that can be expressed as

$$E_2 = \alpha C(s) + \beta C(2s) + \gamma C(4s).$$

What are the three coefficients α, β, and γ?

2. Suppose that a finite-difference computation for some quantity C has given the following results:

h	C
1.0000	26.8285
0.5000	16.2720
0.2500	12.6726
0.1250	11.1832
0.0625	10.5062

(a) What is your estimate of the true value of C?
(b) How accurate do you think your value is?

COMPUTER PROJECT 1-1

Write and run a computer program to evaluate numerically the integrand

$$I = \int_{-1}^{+1} \frac{1}{\sqrt{1 - x^2}}\, dx.$$

Use an approach similar to that used in the text. Take advantage of symmetry and use Richardson extrapolation. Compare your results with the analytic solution.

REFERENCES

1. K. E. Atkinson, *An Introduction to Numerical Analysis* (Second Edition), John Wiley & Sons, New York, 1989.
2. R. L. Burdem and J. D. Faires, *Numerical Analysis* (Fourth Edition), PWS-Kent Publishing Company, Boston, 1989.
3. C. F. Gerald and P. O. Wheatley, *Applied Numerical Analysis* (Fourth Edition), Addison-Wesley Publishing Company, Reading, Mass., 1989.
4. L. Lapidus and G. F. Pinder, *Numerical Solution of Partial Differential Equations in Science and Engineering*, John Wiley & Sons, New York, 1982.
5. Germund Dahlquist and Åke Björck, *Numerical Methods*, Prentice Hall, Englewood Cliffs, N.J., 1974.
6. T. Itoh (ed.), *Planar Transmission Line Structures*, IEEE Press, New York, 1987.
7. T. Itoh (ed.), *Numerical Techniques for Microwave and Millimeter-Wave Passive Structures*, John Wiley & Sons, New York, 1989.
8. S. R. H. Hoole, *Computer-Aided Analysis and Design of Electromagnetic Devices*, Elsevier, New York, 1989.
9. Eikichi Yamashita (ed.), *Analysis Methods for Electromagnetic Wave Problems*, Artech House, Norwood, Mass., 1990.
10. R. Mittra, *Computer Techniques for Electromagnetics*, Hemisphere Publishing Corporation, New York, 1973.
11. Z. J. Cendes (ed.), *Computational Electromagnetics*, North-Holland, Amsterdam, 1986.
12. Gene H. Golub and Charles F. Van Loan, *Matrix Computations*, The Johns Hopkins University Press, Baltimore, 1983.
13. I. S. Duff, A. M. Erisman, and J. K. Reid, *Direct Methods for Sparse Matrices*, Oxford University Press, Oxford, 1986.
14. W. H. Press, B. P. Flannery, S. A. Teukolsky, and W. T. Vetterling, *Numerical Recipes: The Art of Scientific Computing* (*The FORTRAN Version*), Cambridge University Press, Cambridge, 1989.
15. G. I. Marchuk and V. V. Shaidurov, *Difference Methods and Their Extrapolations*, Springer-Verlag, New York, 1983.

CHAPTER TWO

Finite-Difference Method

2.1. FINITE DIFFERENCES IN ONE DIMENSION

The first method we consider for solving differential equations is probably the simplest, the finite-difference method [1, 2]. The first step in applying the finite-difference method is to select a discrete set of values of x, the mesh points, as illustrated in Fig. 2-1 for one dimension. If the spatial interval that separates the points in the mesh is denoted by h, the mesh points can be denoted by the discrete index k, where

$$x_k = kh. \tag{2-1}$$

One then concentrates on the values of the function f at the mesh points and in a sense ignores the values at other points. To simplify the notation, one associates with the function f of the continuous variable x, the function F of the discrete index k, defined by

$$F(k) = f(kh). \tag{2-2}$$

The next step in the solution is to replace the differential equation, which involves derivatives of a function of the continuous variable x, by algebraic expressions involving the function of the discrete index k. First, we need approximations for the derivatives.

-1 0 1 2

FIGURE 2-1. Mesh of points in one dimension.

Consider the Taylor expansion

$$f(x+h) = f(x) + hf'(x) + \frac{h^2}{2}f''(x) + \cdots, \tag{2-3}$$

from which

$$\frac{f(x+h) - f(x)}{h} = f'(x) + \frac{h}{2}f''(x) + \cdots. \tag{2-4}$$

The left side is an approximation to the derivative $f'(x)$, but as h approaches 0, the error approaches 0 only as the first power of h and thus this is not a desirable approximation. The left side of (2-4) is a better approximation to the derivative at the midpoint. To see this, consider the expansions

$$f(x+h) = f\left(x + \frac{h}{2}\right) + \frac{h}{2}f'\left(x + \frac{h}{2}\right) + \frac{h^2}{8}f''\left(x + \frac{h}{2}\right) + \frac{h^3}{48}f'''\left(x + \frac{h}{2}\right) - \cdots \tag{2-5}$$

and

$$f(x) = f\left(x + \frac{h}{2}\right) - \frac{h}{2}f'\left(x + \frac{h}{2}\right) + \frac{h^2}{8}f''\left(x + \frac{h}{2}\right) - \frac{h^3}{48}f'''\left(x + \frac{h}{2}\right) + \cdots \tag{2-6}$$

Subtraction of these expansions yields

$$\frac{f(x+h) - f(x)}{h} = f'\left(x + \frac{h}{2}\right) + \frac{h^2}{24}f'''\left(x + \frac{h}{2}\right) + \cdots. \tag{2-7}$$

Note that the error decreases as the second power of h and hence this is a more desirable approximation. This is a good approximation for the first derivative between nodal points, but usually we need approximations at the nodal points. To find such an approximation, combine the expansion

$$f(x-h) = f(x) - hf'(x) + \frac{h^2}{2}f''(x) - \frac{h^3}{6}f'''(x) + \cdots \tag{2-8}$$

with (2-3) to get

$$\frac{f(x+h) - f(x-h)}{2h} = f'(x) + \frac{h^2}{6}f'''(x) + \cdots. \tag{2-9}$$

This is now the sort of approximation that we wish to use. In a similar manner, one can show that

$$\frac{f(x+h) - 2f(x) + f(x-h)}{h^2} = f''(x) + \frac{h^2}{12}f''''(x) + \cdots, \tag{2-10}$$

and the left side of this equation is a good approximation for the second derivative. In general we prefer to use approximations such as (2-9) and (2-10), referred to as central difference formulas, because of the greater accuracy.

Several generalizations are possible. The left side of (2-10) involves what is known as a three-point formula. Even more accurate approximations involving five and more values are available, but it is questionable whether the greater accuracy is worth the increased programming time, and we shall not use them. Approximations for higher-order derivatives can be derived using the same approach, but approximations for first and second derivatives are what we need for the electromagnetic problems studied here. One generalization that we will encounter later in this chapter involves approximations for partial derivatives when we study problems in two and three dimensions.

2.2. ONE-DIMENSIONAL DIFFERENTIAL EQUATION EXAMPLE

As a specific example, consider the equation

$$\frac{d^2f}{dx^2} + 4f = 0 \qquad \text{for } 0 \le x \le 1 \tag{2-11}$$

with the boundary conditions

$$f(0) = 0 \qquad \text{and} \qquad f(1) = 1. \tag{2-12}$$

Suppose that we wish to use finite-difference approximations to find $f(0.5)$. The largest value of h that we can use is 0.5. As shown in Fig. 2-2, there are three nodal values, of which only one, $F(1)$, is unknown. If we replace the derivative in (2-11) by the three-point central difference formula centered at $k = 1$, the differential equation becomes

$$\frac{F(0) - 2F(1) + F(2)}{(0.5)^2} + 4F(1) = 0 \tag{2-13}$$

and the boundary conditions lead to

$$F(0) = 0 \qquad \text{and} \qquad F(2) = 1. \tag{2-14}$$

Solution for $F(1)$ gives

$$F(1) = \frac{F(0) + F(2)}{1}. \tag{2-15}$$

Insertion of the boundary values from (2-14) into (2-15) and evaluation of $F(1)$ leads to

$$F(1) = 1.0000. \tag{2-16}$$

0	1	2	k
0	0.5	1	x

FIGURE 2-2. Three-point mesh for differential equation.

0	1	2	3	4	k
0		0.5		1.0	x

FIGURE 2-3. Five-point mesh for differential equation.

This gives us a first value for $f(0.5)$ of 1.0000. For a more accurate value, take h as

$$h = 0.25, \tag{2-17}$$

and then we have the situation shown in Fig. 2-3, with two external mesh points at which the functional values are determined by the boundary conditions, namely

$$F(0) = 0 \quad \text{and} \quad F(4) = 1 \tag{2-18}$$

and three internal points at which the functional values are to be determined. Use of the central-difference formula (2-10) at each of the internal mesh points leads to the three equations

$$\begin{aligned}
\frac{F(0) - 2F(1) + F(2)}{0.0625} + 4F(1) &= 0 \\
\frac{F(1) - 2F(2) + F(3)}{0.0625} + 4F(2) &= 0 \\
\frac{F(2) - 2F(3) + F(4)}{0.0625} + 4F(3) &= 0,
\end{aligned} \tag{2-19}$$

which can be rearranged as

$$\begin{aligned}
1.75F(1) - F(2) &= F(0) \\
-F(1) + 1.75F(2) - F(3) &= 0 \\
-F(2) + 1.75F(3) &= F(4).
\end{aligned} \tag{2-20}$$

Substitution of the boundary conditions and direct solution of these equations gives

$$\begin{aligned} F(1) &= 0.5378 \\ F(2) &= 0.9412 \\ F(3) &= 1.1092. \end{aligned} \tag{2-21}$$

The value for $F(2)$ is an improved value for $f(0.5)$. We have calculated $f(0.5)$ for two values of h, 0.5 and 0.25. We can continue the sequence of calculations by using for h, the values

$$h = \frac{0.5}{M} \qquad \text{for } M = 1, 2, 4, \ldots, \tag{2-22}$$

where we have already used $M = 1$ and 2. The values of $f(0.5)$ that result are shown in the following table.

M	$f(0.5)$
1	1.0000
2	0.9412
4	0.9292
8	0.9263
16	0.9256
32	0.9255

Analytical solution of the differential equation gives 0.9254.

2.3. FINITE DIFFERENCES IN TWO DIMENSIONS

As we just illustrated in the one-dimensional example, the first step in the finite-difference method is to establish a mesh of points. Figure 2-4 illustrates a mesh in two dimensions. As in that illustration, a uniform mesh is usually used with simple problems. Functions of discrete indices are then defined by

$$F(i, j) = f(ih, jh). \tag{2-23}$$

If the complexity of the field varies over the region of interest, a

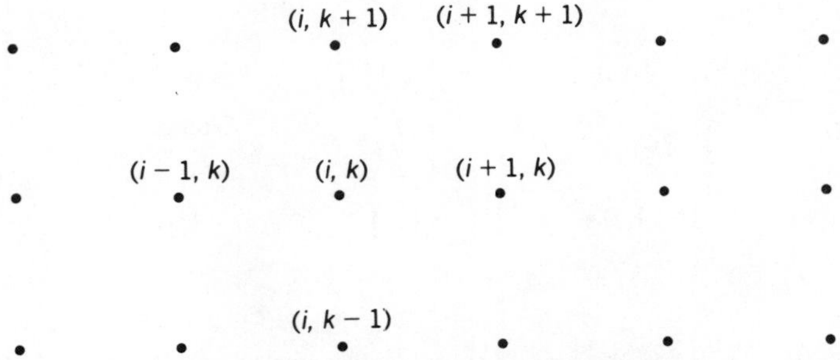

(i, k + 1) (i + 1, k + 1)

(i − 1, k) (i, k) (i + 1, k)

(i, k − 1)

FIGURE 2-4. Two-dimensional mesh.

nonuniform mesh may be desirable. The derivative approximations developed for one dimension can easily be generalized to two and three dimensions. For example, the finite-difference approximation for the Laplacian of a function of two variables is

$$\frac{\partial^2 f}{\partial x^2} + \frac{\partial^2 f}{\partial x^2} = \frac{\begin{array}{l} f(x+h, y) + f(x-h, y) \\ +f(x, y+h) + f(x, y-h) - 4f(x, y) \end{array}}{h^2}, \quad (2\text{-}24)$$

which in discrete terms is

$$\nabla^2(i, j) = \frac{\begin{array}{l} F(i-1, j) + F(i+1, j) \\ +F(i, j-1) + F(i, j+1) - 4F(i, j) \end{array}}{h^2}. \quad (2\text{-}25)$$

Use of this approximation with Laplace's equation

$$\frac{\partial^2 f}{\partial x^2} + \frac{\partial^2 f}{\partial y^2} = 0, \quad (2\text{-}26)$$

for example, results in the discrete equation

$$\begin{aligned} F(i-1, j) + F(i+1, j) + F(i, j-1) \\ +F(i, j+1) - 4F(i, j) = 0. \quad (2\text{-}27) \end{aligned}$$

If one such equation is written for each unknown value of F, a set

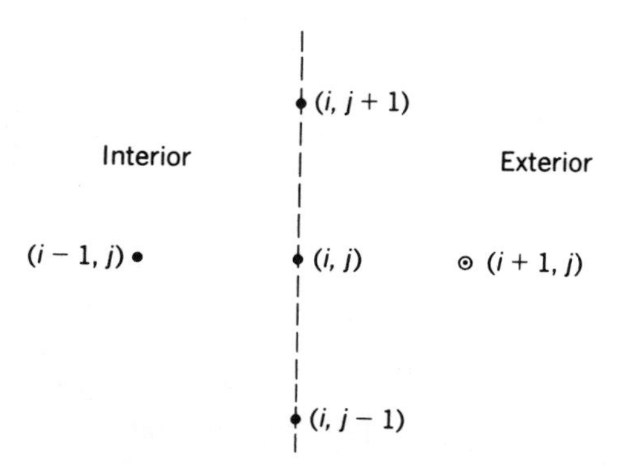

FIGURE 2-5. Neumann boundary condition.

of linear algebraic equations results. If the function f is specified on the boundary, a boundary condition known as the *Dirichlet condition*, the value of F is specified on all the boundary points. With this condition we wish to satisfy (2-27) at each internal point. If the internal point at which we wish to satisfy (2-27) adjoins a boundary point, some of the functional values in this equation are known. Another frequently encountered boundary condition is one known as the *Neumann condition*, where the normal derivative is specified and frequently to the value zero. For the situation of Fig. 2-5, the point (i, j) is a boundary point where the Neumann condition

$$\frac{\partial f}{\partial x} = 0 \tag{2-28}$$

is to be satisfied. Because the functional value $F(i, j)$ is unknown, we must use (2-27) at this point. This requires a value for $F(i + 1, j)$, but the point $(i + 1, j)$ is outside the computation region and thus the functional value $F(i + 1, j)$ is not available. The boundary condition (2-28) leads to

$$F(i + 1, j) = F(i - 1, j) \tag{2.29}$$

and we could use this relation to replace the value for the point outside the computation region. Then (2-27) would be replaced by

$$2F(i-1,j) + F(i,j+1) + F(i,j-1) - 4F(i,j) = 0. \quad (2\text{-}30)$$

The set of linear algebraic equations can be solved directly, but unlike the one-dimensional example, even moderately small values of h may result in so many unknowns that direct solution is not feasible. Iterative, or indirect, methods have been found to be preferable for most two- and three-dimensional problems. For such a solution, write (2-27) in the form

$$F(i,j) = \frac{F(i+1,j) + F(i-1,j) + F(i,j+1) + F(i,j-1)}{4}. \quad (2\text{-}31)$$

The process of iteration starts with some initial values $F^0(i,j)$ for the unknown values, and cycles repeatedly through all (i,j) pairs that correspond to unknown values of F, and uses (2-31) in the form

$$F^1(i,j) = \frac{F^0(i+1,j) + F^0(i-1,j) + F^0(i,j+1) + F^0(i,j-1)}{4}. \quad (2\text{-}32)$$

can be used to calculate an improved set of values $F^1(i,j)$. Then a second set of values can be calculated, and so on, using the general formula

$$F^n(i,j) = \frac{\begin{array}{c} F^{n-1}(i+1,j) + F^{n-1}(i-1,j) \\ + F^{n-1}(i,j+1) + F^{n}(i,j-1) \end{array}}{4}, \quad (2\text{-}33)$$

where, as before, the superscript indicates the number of the iteration. This process can be continued until the solution has converged to suitable accuracy. Because the change from one step of the iteration to the next may involve slight changes in the functional values, convergence is difficult to detect by examining successive steps. A safer procedure is to double the number of iterations successively and examine the changes after 1, 2, 4, 8, and so on, steps of iteration. When these numbers change little, it is safe to assume that the process has converged.

The process of iterating b repeatedly using the same equations for the nodal values is known as *relaxation*, from early structural applications in which the process was envisioned as "relaxing" stress. An improvement that speeds the convergence process is known as *over relaxation*, and for the example of Laplace's equation replaces (2-33) by

$$F^n(i,j) = F^{n-1}(i,j) + R\left[\frac{F^{n-1}(i+1,j) + F^{n-1}(i-1,j) + F^{n-1}(i,j+1) + F^{n-1}(i,j-1)}{4} - F^{n-1}(i,j)\right]. \tag{2-34}$$

If the constant R, known as the relaxation parameter, is equal to unity, the regular relaxation process results. Values of R greater than unity speed convergence, but too large a value of R results in numerical instability. For Laplace's equation, a value of 2 gives instability, and values near 1.5 are shown to give good results. The program shown later uses a value of 1.4.

In calculating the functional values for one iteration step, we used only the values from the preceding step, which is an iterative process known as the *Jacobi process*. For the first calculation in each step that is all we can do. For the second calculation, we have a choice, because the first calculation has given us a more accurate value for one of the numbers. The iterative process that always uses the most recent numbers available is known as the *Gauss–Seidel method*. The Gauss–Seidel method can be shown to converge faster for a wide range of problems, and in addition, is easier to program since we do not need to store more than one value for each unknown.

The initial set of values $F^0(i,j)$ required to start the iterative process can be chosen arbitrarily as all zero or all unity. Convergence to the desired solution occurs faster if one can start the process with a set of initial values that are reasonably close to the final values.

2.4. A TWO-DIMENSIONAL CAPACITANCE EXAMPLE

A simple two-dimensional example is the quasistatic analysis of a shielded parallel plate transmission line, the cross section of which is shown in Fig. 2-6. The potential satisfies Laplace's equation

$$\frac{\partial^2 \phi}{\partial x^2} + \frac{\partial^2 \phi}{\partial y^2} = 0. \tag{2-35}$$

For convenience, we will take the potentials on the upper and lower plates as $+1$ and -1, respectively, and the potential on the outer shield as 0. To minimize the computational effort, we can take advantage of the top-to-bottom antisymmetry and the left-to-right symmetry to replace the situation by the one shown in Fig. 2-7. The

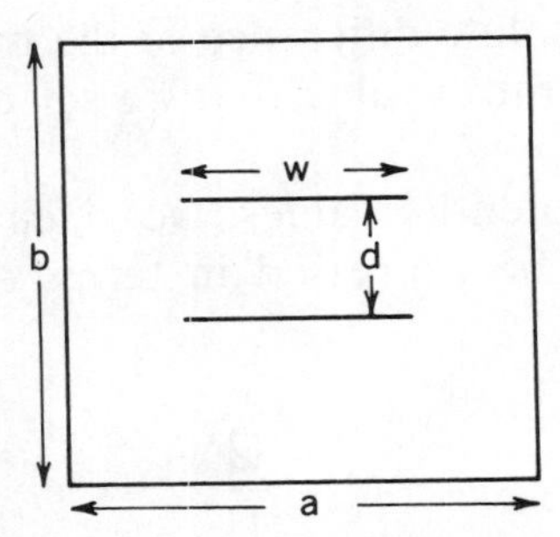

FIGURE 2-6. Cross section of shielded parallel plate transmission line.

FIGURE 2-7. One-fourth of cross section.

methods discussed in the preceding section can be used to solve for the potential $\phi(x, y)$ and then in turn we can calculate the capacitance. The fundamental steps in a computer algorithm to carry out this computation are:

1. Establish parameter values.
2. Establish boundary conditions and initial values.
3. Iterate to determine the potential values.
4. Calculate the capacitance.

We need first to select values for the parameters a, b, w, and d. Since the capacitance is not affected by scaling, we will use instead the three normalized parameters a/d, b/d, and w/d. Then we establish the potential on the shield of 0 volts and on the center conductor of 1 volt. Next we perform a set of iterations. To perform one iteration, we raster scan through all points at which the potential is unknown and at each point utilize the overrelaxation expression. As discussed in the preceding section, a good approach is to examine the results after one iteration, two iterations, four iterations, and so on. After each such set of iterations, we calculate the capacitance and compare it to the previous value. We stop the iteration after the absolute difference of the two capacitance values is less than some tolerance value that we select as one of the input parameters.

For each set of potential values, we calculate the capacitance. The capacitance can be expressed in terms of the charge on the center conductor as

$$C = \frac{Q}{V}, \tag{2-36}$$

where the voltage V is the difference between the potential values on the two conductors. Because this potential difference is 2 volts,

$$C = \frac{Q}{2 \text{ volts}}. \tag{2-37}$$

The charge Q can be calculated from the two-dimensional version of Gauss's law as

$$Q = \epsilon_0 \int E_{\text{out}}\, dl, \tag{2-38}$$

where the line integral surrounds the center conductor and the subscript on E means the normal component away from the center conductor. Because

$$E_{\text{out}} = -E_{\text{in}} = \left(\frac{\partial \phi}{\partial n}\right)_{\text{in}}, \tag{2-39}$$

the integral can be computed as

$$Q = \epsilon_0 \int \left(\frac{\partial \phi}{\partial n}\right)_{\text{in}} dl, \tag{2-40}$$

and thus

$$C = \frac{\epsilon_0}{2 \text{ volts}} \int \left(\frac{\partial \phi}{\partial n}\right)_{\text{in}} dl. \tag{2-41}$$

Since we are analyzing just part of the total region, we will evaluate this line integral for that subregion and multiply the result by 2.

To complete the algorithm, we need to select values of the finite-difference distance h and select a criterion for stopping the iteration. The largest value that we can take for h is

$$h_m = \frac{d}{2} \tag{2-42}$$

assuming that w is a multiple of d, and we will select a set of values defined by

$$h = \frac{h_m}{M} = \frac{d}{2M}, \tag{2-43}$$

where M is an integer whose value we shall, in succession, take as $1, 2, 4, \ldots$ up to as many values as we wish to use.

We thus calculate a value of capacitance for each value of h. The final step is to apply Richardson extrapolation, as discussed in Chapter 1, to these capacitance values to arrive at our best estimate. A FORTRAN program that implements the complete algorithm follows.

FORTRAN Program for the Capacitance of a Shielded Parallel-Plate Transmission Line

```
      program shcapfd
C     capacitance of shielded capacitor
C     calculated with finite differences
C     line integral of E field
C     solution by iteration
      real add,bdd,R, delta,cap,old
      real z
      integer C,W,WW,M,nx,ny,i,j,k,kk
      real V(200,200),ZZ(10,10),Y(10)
C     input parameters
      print*
      print*
      print*
      print*
      print*,'    shielded capacitor'
      print*,'    finite-difference approximation'
      print*,'    capacitance from line integral'
      print*,'    solution by iteration'
      print*
      print*,'    what is a/d?'
      read*,add
      print*
      print*,'what is b/d?'
      read*,bdd
      print*
      print*,'how many values of M?'
      read*,WW
      print*
```

```
      print*,'   enter tolerance'
      read*,tol
      print*
      print*
      R=1.4
C     print parameters
      print*,'   shielded capacitor'
      print*,'   a/d is ',add
      print*,'   b/d is ',bdd
      print*,'   tolerance is ',tol
      print*
      print*,'     M  original cap  extrapolations'
      print*
C     start calculation
      M=1
      W=1
      do while (W.LE.WW)
      nx=add*M
      ny=bdd*M
C     set initial and boundary conditions
      doj=0,ny
        doi=0,nx
          V(i,j)=0
        end do
      end do
      do j=1,ny-1
        do i=0,nx-1
          V(i,j)=1
        end do
      end do
      cap=1
      old=0
      k=1
      kk=1
      do while (abs(old-cap).GT.tol)
        old=cap
        do while (k.LE.kk)
          doj=1,ny-1
```

```
          doi=0,nx-1
            if(j .EQ. M) then
              if (i .GT. M) then
              z=V(i-1,j)+V(i+1,j)
              z=z+V(i,j-1)+V(i,j+1)
              delta=z/4-V(i,j)
              V(i,j)=V(i,j)+R*delta
            end if
          else
            if (i.EQ.0) then
              z=2*V(i+1,j)+V(i,j-1)
                z=z+V(i,j+1)
                delta=z/4-V(i,j)
                V(i,j)=V(i,j)+R*delta
                else
                  z=V(i-1,j)+V(i+1,j)
                  z=z+V(i,j-1)+V(i,j+1)
                  delta=z/4-V(i,j)
                  V(i,j)=V(i,j)+R*delta
                end if
              end if
            end do
          end do
          k=k+1
        end do
        z=(V(0,1)+V(0,ny-1))/2
        do i=1,nx-1
          z=z+V(i,1)+V(i,ny-1)
        end do
        do j=1,ny-1
        z=z+V(nx-1,j)
      end do
      cap=8.854187*z
      kk=2*kk
    end do
    ZZ(W,1)=cap
C   extrapolation calculation
    C=1
    do j=2,W
```

```
      C=2*C
      ZZ(j,W)=(C*ZZ(j-1,W)-ZZ(j-1,W-1))/(C-1)
    end do
C   print results
    print*,Y(W),ZZ(1,W),ZZ(W,W)
    W=W+1
    M=2*M
   end do
   end
```

The results of a computation run for the parameter values

$$\frac{a}{d} = 2$$
$$\frac{b}{d} = 2 \tag{2-44}$$
$$\frac{w}{d} = 1$$

are shown in the following table.

M	Capacitance	Extrapolations
1	35.41675	35.41675
2	29.93558	24.45442
4	27.83906	26.17192
8	26.89887	26.01054

2.5. OPEN REGIONS

Clearly, use of the finite-difference method to solve problems such as the example in the preceding section is straightforward. The fact that we need to consider only the region inside the shield means that we have only a finite number of mesh points. Situations with open regions, however, present a serious problem, because there are an infinite number of mesh points, which clearly cannot be handled. One approach to such a situation is to restrict the computation to a finite portion of the space and then to successively expand this portion of the space. For example, if we consider the parallel-plate transmission line, we can retain the shield but make the shielded region larger and larger. Richardson extrapolation can then be utilized to estimate the value of capacitance with the shield at an infinite distance. The following table gives the results of such a computation.

$a/d = b/d$	Capacitance (pF/m)	Extrapolations
2	26.01054	26.01054
4	20.18555	14.36056
8	19.07868	19.17556
16	18.81944	18.69644

Although it is not clear from this table how well the value has converged, the approximate value of 18.7 does agree with the value we will derive in a later chapter from the much more efficient moment method. Although this method of handling open regions does work, it leads quickly to very large numbers of points and is not very computationally efficient. The moment method that we will treat later proves to be much better suited to open regions.

2.6. GENERALIZATIONS

The replacement of derivatives by finite-difference approximations clearly can be applied to other partial differential equations. Although we here discussed only one- and two-dimensional problems, the generalization to three dimensions presents no problems in principle. The additional dimension does, however, lead to a larger number of points and increased run time. This is why the tendency is to solve two-dimensional problems wherever possible. Three-dimensional problems will be discussed further in Chapter 4, where time-domain problems are discussed.

Because the finite-difference method is based on a rectangular mesh, rectangular conditions on rectangular boundaries can be satisfied in a natural way. When the boundary is curved, special modifications must be made. One way is to use straight "stair-step"

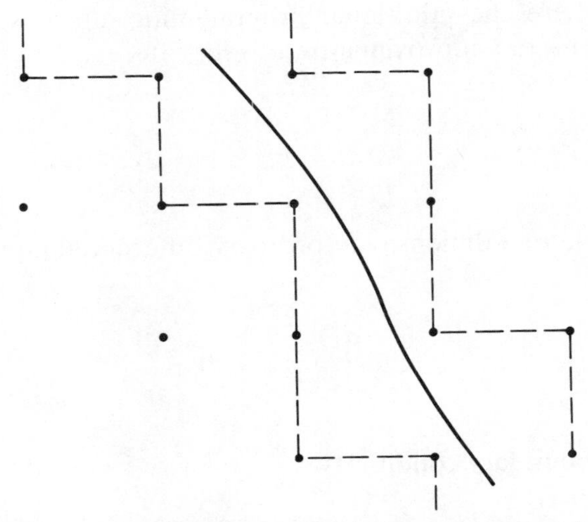

FIGURE 2-8. Internal and external stair-step approximation to curved boundary.

approximations to the boundary. One can either fit the straight-line boundary approximation inside or outside the true boundary, as shown in Fig. 2-8. One can solve the problem twice, once with an inside approximation and once with an outside approximation, and then average the results. A more sophisticated way to accommodate curved boundaries is to modify the finite-difference approximation to derivatives to allow unequal spacing whenever a nearest point falls on the boundary, as shown in Fig. 2-9. Although there may be

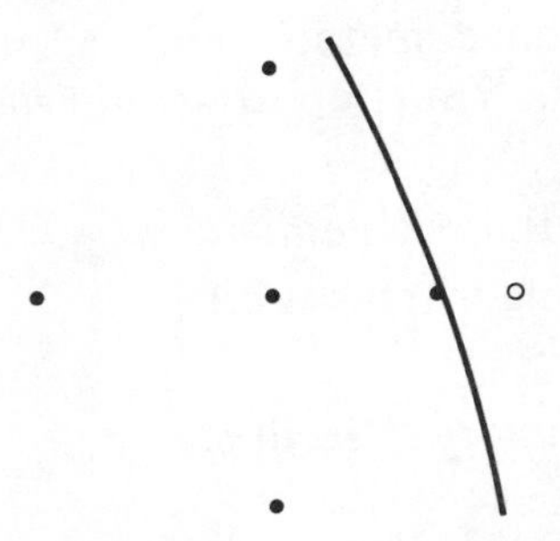

FIGURE 2-9. Unequal spacing for curved boundary nodes.

problems where the additional programming effort is warranted, usually a stair-step approximation is adequate.

PROBLEMS

1. The simple one-dimensional ordinary differential equation

$$\frac{d^2f}{dx^2} + f = 0$$

with the boundary conditions

$$f(0) = 0 \qquad \text{and} \qquad f(1) = 1$$

is to be solved with the finite-difference method. For a very crude mesh, take $h = 0.5$. What is $f(0.5)$?

2. For the values of ϕ at the nonuniformly spaced points in the figure, calculate an approximate value for $\partial^2\phi/\partial x^2 + \partial^2\phi/\partial y^2$.

COMPUTER PROJECT 2-1

This computer project is the calculation of the capacitance per unit length for the square coaxial transmission line shown in the figure. Use the finite-difference method and calculate the value to an accuracy of 0.1 pF/m. Your report should include:

a. A statement of the problem.
b. Equations and algorithms used.
c. A program listing.
d. Intermediate numerical results.
e. The answer.

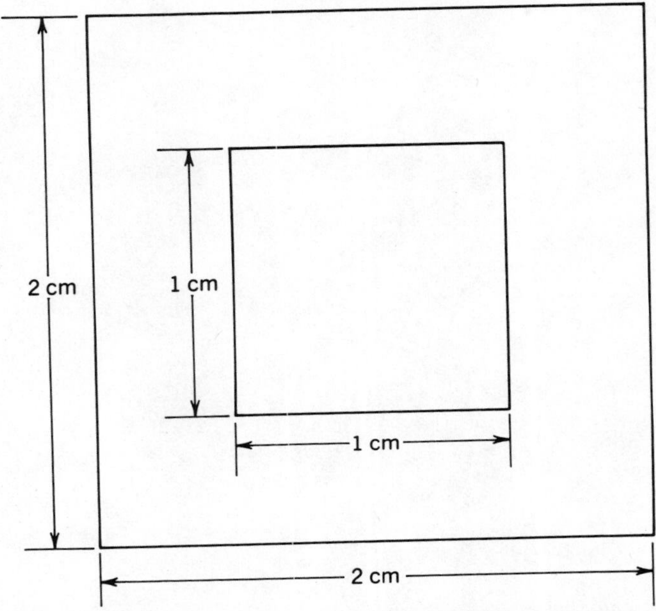

COMPUTER PROJECT 2-1. Cross section of sqaure coaxial-like transmission line.

REFERENCES

1. G. D. Smith, *Numerical Solution of Partial Differential Equations: Finite Difference Methods* (Third Edition), Oxford University Press, Oxford, 1985.
2. John C. Strikwerda, *Finite Difference Schemes and Partial Differential Equations*, Wadsworth & Brooks, Belmont, Calif., 1989.

CHAPTER THREE

Finite-Difference Determination of Eigenvalues

3.1. EIGENVALUES IN ONE DIMENSION

An extremely important problem in electromagnetics is the determination of resonant frequencies of a structure such as a cavity. Mathematically, this is described as an *eigenvalue problem*. A simple eigenvalue problem in one dimension involves determining the values of k that are consistent with nontrivial solutions of the differential equation for $f(x)$:

$$\frac{d^2 f}{dx^2} + k^2 f = 0 \quad \text{with} \quad f(0) = 0 \quad \text{and} \quad f(1) = 0. \tag{3-1}$$

The general solution of the equation is

$$f(x) = A \cos kx + B \sin kx, \tag{3-2}$$

and the imposition of the boundary conditions requires $A = 0$ and $k = n\pi$, where n is an integer. One way to find the values of k numerically is to use the finite-difference approximation for the second derivative, which leads to the set of equations

$$\frac{f_{i-1} - 2f_i + f_{i+1}}{h^2} + k^2 f_i = 0, \tag{3-3}$$

which can be written in the form

$$-f_{i-1} + \alpha f_i - f_{i+1} = 0, \tag{3-4}$$

where

$$\alpha = 2 - h^2 k^2. \tag{3-5}$$

Nontrivial solutions of these algebraic equations exist only if the determinant of the system of equations is zero. When the number of unknown functional values is small, the determinant can be evaluated explicitly. For example, if $h = 0.25$, there are three unknowns, and setting the determinant equal to zero gives

$$\begin{vmatrix} \alpha & -1 & 0 \\ -1 & \alpha & -1 \\ 0 & -1 & \alpha \end{vmatrix} = 0. \tag{3-6}$$

Evaluation of the determinant leads to the equation

$$\alpha(\alpha^2 - 2) = 0, \tag{3-7}$$

which has the solutions

$$\alpha = 0, \pm\sqrt{2} \tag{3-8}$$

which leads to k values of

$$k = 3.0615, 5.6569, \text{and } 7.3910. \tag{3-9}$$

Compared to the true solutions of 3.1416, 6.2832, and 9.4248, the first value is a fairly good approximation, the second is so-so, and the third is not a very good approximation.

3.2. WAVEGUIDE-MODE EXAMPLE

A more realistic example is the problem of finding in two or three dimensions the eigenvalues that determine the propagation parameters of waveguide modes, resonant frequencies of resonators, and many other physical parameters. As a simple example, consider a rectangular waveguide, with propagation in the z-direction. For the TM mode, the four field components E_x, E_y, H_x, and H_y can be expressed in terms of E_z. In turn, E_z can be written as

$$E_z(x, y, z, t) = E(x, y)e^{j(\omega t - \beta z)}, \qquad (3\text{-}10)$$

where E satisfies

$$\frac{\partial^2 E}{\partial x^2} + \frac{\partial^2 E}{\partial y^2} + k^2 E = 0 \qquad (3\text{-}11)$$

and

$$E = 0 \qquad (3\text{-}12)$$

on the boundary. The parameter k determines the phase parameter β through

$$k^2 = \omega^2 \mu \epsilon - \beta^2. \qquad (3\text{-}13)$$

The eigenvalue problem is to solve for $E(x, y)$ and k simultaneously so that (3-11) is satisfied.

To use the finite-difference method, a mesh of points is established, separated by the uniform interval of h. With the previously derived approximation for the Laplacian, the discrete equation that results is

$$\frac{E(i+1, j) + E(i-1, j) + E(i, j+1) + E(i, j-1) - 4E(i, j)}{h^2} + k^2 E(i, j) = 0, \qquad (3\text{-}14)$$

which can be written as

$$-E(i+1, j) - E(i-1, j) - E(i, j+1) - E(i, j-1) + (4 - k^2 h^2)E(i, j) = 0. \qquad (3\text{-}15)$$

With the notation

$$\alpha = 4 - k^2h^2, \tag{3-16}$$

this equation is

$$-E(i+1,j) - E(i-1,j) - E(i,j+1) - E(i,j-1) + \alpha E(i,j) = 0. \tag{3-17}$$

This is a set of linear algebraic equations, one for each interior point. These equations can be written in matrix form as

$$A(\alpha)E = 0, \tag{3-18}$$

where the dependence of the matrix coefficients on α is explicitly written. For the set of equations (3-18) to have nonzero solutions for E, the determinant of the matrix must satisfy

$$\det[A(\alpha)] = 0. \tag{3-19}$$

As is well known, this determinant is a polynomial in α with degree equal to the number of points. For

$$h = \frac{a}{4}, \tag{3-20}$$

the situation in Fig. 3-1 results. There are three unknown values

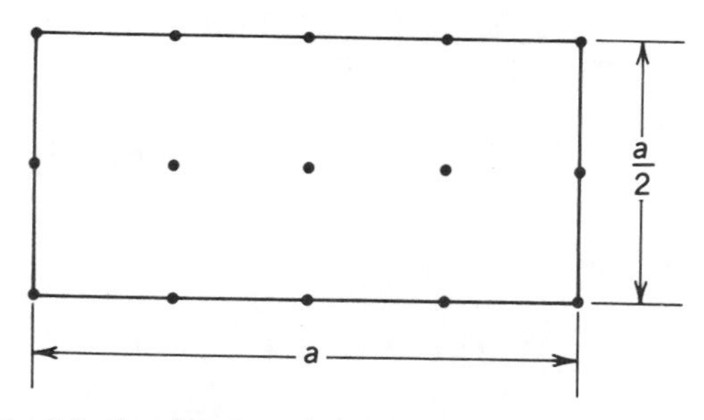

FIGURE 3-1. Mesh of points for finite-differences with $h = a/4$.

and the equations (3-18) are

$$\begin{aligned} \alpha E_1 - E_2 &= 0 \\ -E_1 + \alpha E_2 - E_3 &= 0 \\ E_2 + \alpha E_3 &= 0. \end{aligned} \tag{3-21}$$

The determinant of this set of equations, which happens to be the same determinant as that considered in the preceding section, satisfies

$$\begin{vmatrix} \alpha & -1 & 0 \\ -1 & \alpha & -1 \\ 0 & -1 & \alpha \end{vmatrix} = 0, \tag{3-22}$$

which leads to the polynomial equation

$$\alpha^3 - 2\alpha = 0. \tag{3-23}$$

This polynomial has the zeros $-\sqrt{2}$, 0, and $\sqrt{2}$, which lead to values for ka of 6.4322, 8.0000, and 9.3074. These are approximations to the first three eigenvalues. To get more accurate values, we need of course to use smaller values of h. A value for h of

$$h = \frac{a}{8}, \tag{3-24}$$

for example, leads to the situation shown in Fig. 3-2, which shows that there are 21 internal points. To write explicitly the determinant that corresponds to (3-22) as a polynomial in α is not an attractive

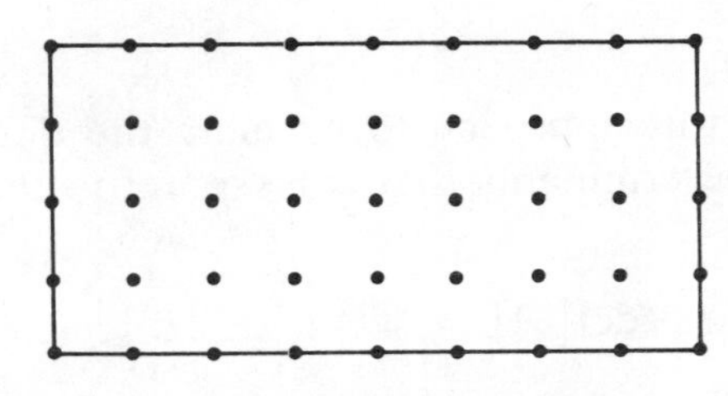

FIGURE 3-2. Mesh of points for finite-differences with $h = a/8$.

task and for even smaller values of h almost impossible. A number of algebraic techniques [1] have been devised to find eigenvalues of such determinants. In the next two sections we consider two numerical techniques.

3.3. NUMERICAL EVALUATION OF THE DETERMINANT

One approach is to use an algorithm that numerically evaluates the determinant [2]. First, we need to count and number the internal points. Then we need to evaluate the matrix coefficients that correspond to (3-19). Evaluation of the determinant can be carried out using a modification of an algorithm for matrix inversion known as *LU factorization*. If the matrix A can be factored as

$$A = LU, \tag{3-25}$$

where L is a lower diagonal matrix with zeros for all elements above the diagonal and U is an upper matrix with zeros for all elements below the diagonal. The principal use of this approach is the solution of the matrix equation

$$AX = B \tag{3-26}$$

by writing it as

$$LUX = B \tag{3-27}$$

and then writing this as the two equations

$$LY = B \tag{3-28}$$

and

$$UX = Y. \tag{3-29}$$

We wish to use this approach to evaluate the determinant of the matrix A. The determinants that correspond to these matrices are related by

$$\det[A] = \det[L]\det[U]. \tag{3-30}$$

One factorization algorithm, known as *Crout's algorithm*, results in the matrix U having ones for the diagonal terms and since the determinant of a diagonal matrix is the product of the diagonal terms, the determinant of U is unity. Thus, with this algorithm,

$$\det[A] = \det[L], \tag{3-31}$$

and since L also is a diagonal matrix, its determinant is the product of its diagonal terms, and thus

$$\det[A] = \prod_{j=1}^{N} L(i,i). \tag{3-32}$$

Although unlikely, one of the values of ka chosen could be a zero of the determinant (and in the first example we will soon consider, this happens). Provision must be included to recognize this possibility and avoid trying to divide by zero in the factorization process. A FORTRAN program for the TM modes of a rectangular waveguide follows.

```
      program tmdet
      integer M,R,RR(100,100),W,F
      real alpha,ka,kh,A(110,110),minka,maxka
      integer i,j,k,N,imax,indx(110)
      real VV(110)
      real aamax, sum, dum, D
C     input parameters
      print*,'what is M?'
      read*,M
      print*,'what is minimum ka?'
      read*,minka
      print*,'what is maximum ka?'
      read*,maxka
C     count and number internal points
      R=0
      doj=1,2*M-1
        doi=1,4*M-1
          R=R+1
          RR(i,j)=R
        end do
      end do
      N=R
      print*,'number of internal points is',N
```

```
C     print column headings
      print*
      print*,'    ka        sign(det)'
      print*
C     calculate values of ka and alpha
      do W=0,10
        ka=minka+W*(maxka-minka)/10.
        kh=ka/4./M
        alpha=4.-kh*kh
      do j=1,N
          do i=1,N
              A(i,j)=0.
          end do
        end do
C       evaluate matrix elements
        do j=1,2*M-1
          do i=1,4*M-1
              A(RR(i,j),RR(i,J))=alpha
              A(RR(i,j),RR(i-1,j))=-1.
              A(RR(i,j),RR(i+1,j))=-1.
              A(RR(i,j),RR(i,J-1))=-1.
              A(RR(i,j),RR(i,j+1))=-1.
              if (j.EQ.1) then
                   A(RR(i,j),RR(i,J-1))=0.
              else if (j.EQ.2*M-1) then
                   A(RR(i,j),RR(i,j+1))=0.
              end if
              if (i.EQ.1) then
                   A(RR(i,j),RR(i-1,j))=-1.
              else if (i.EQ.4*M-1) then
                   A(RR(i,j),RR(i+1,j))=-1.
              end if
          end do
        end do
C       start LU decomposition
        D=1.
        F=1
        do i=1,N
          vv(i)=1.
        end do
```

```
do i=1,N
  aamax=0.
  do j=1,N
  if (abs(A(i,j)).GT.aamax) aamax=
  abs(A(i,j))
  end do
  if (aamax.EQ.0.) then
      F=0
  else
      vv(i)=1./aamax
  end if
end do
do j=1,N
  do i=1,j-1
      sum=A(i,j)
      do k=1,i-1
         sum=sum-A(i,k)*A(k,j)
      end do
      A(i,j)=sum
  end do
  aamax=0
  do i=j,N
      sum=A(i,j)
      do k=1,j-1
         sum=sum-A(i,k)*A(k,j)
      end do
      A(i,j)=sum
      dum=vv(i)*abs(sum)
      if (vv(i).GT.aamax) then
         imax=i
         aamax=dum
      end if
  end do
  if (j.NE.imax) then
      do k=1,N
         dum=A(imax,k)
         A(imax,k)=A(j,k)
         A(j,k)=dum
      end do
      D=-D
```

```
                  vv(imax)=vv(j)
              end if
              indx(j)=imax
              if (A(j,j).EQ.0) then
                  F=0
              else
                  if (j.NE.N) then
                      dum=1./A(j,j)
                      do i=j+1,N
                         A(i,j)=A(i,j)*dum
                      end do
                    end if
              end if
           end do
C          calculate sign of determinant
           if (F.EQ.0) then
             D=0.
           else
             do j=1,N
                  D=D*A(j,j)/abs(A(j,j))
             end do
           end if
C          print solution for one value of ka
           print *,ka,D
C        repeat calculation for another value of ka
         end do
         end
```

An algorithm that evaluates the determinant can be used in several ways. The determinant is a function of α, or equivalently of ka, and the Newton–Raphson method for finding zeros of functions can be combined with the determinant evaluation algorithm. The bisection method is another useful way to find zeros of functions. For interactive computations, a scheme based on decimal division seem to more useful. The determinant is evaluated at 11 values of ka, say from 0 to 10, and one looks for successive values of ka for which the sign of the determinant changes sign. If this process is to be followed, only the sign of the determinant must be calculated and not the value. This simplifies numerical problems that may arise with large matrices. Then one repeats the process for another 11 values, spanning the region where the determinant changed sign in the first calculation. In this way, a new decimal digit is determined, and the process can be continued until the accuracy desired is achieved. As an example, for $M = 2$, the following results are obtained.

ka	sign(det)
0	−1
1	−1
2	−1
3	−1
4	−1
6	−1
7	+1
8	0
9	−1
10	+1

One can see that there is a zero between 6 and 7, one at 8 exactly, and one between 9 and 10. If we desire to get a more accurate value of the first one, for example, we compute *ka* between 6 and 7 and get the following results.

ka	sign(det)
6.0	−1
6.1	−1
6.2	−1
6.3	−1
6.4	−1
6.5	+1
6.6	+1
6.7	+1
6.8	+1
6.9	+1
7.0	+1

The zero is seen to be between 6.4 and 6.5. After several more such calculations, we get a value of approximately 6.43215. For three values of M, the first three zeros are given in the following table. The table also shows the extrapolated value and the analytic result.

M				
1	2	4	Extrapolation	Analytic Result
6.43215	6.87265	6.98655	7.02955	7.02481
8.00000	8.65915	8.82875	8.89168	8.88577
9.30735	10.79385	11.19205	11.36022	11.32717

This entire process could be programmed to completely automate calculation of the eigenvalues. Although in principle this approach should work as well for the TE modes as it does for the TM modes, unfortunately some of the zeros of the determinant are multiple and are extremely difficult to locate.

A number of other matrix-based algorithms have been developed for calculation of eigenvalues. The well-known power method is an iterative process that converges to the largest eigenvalue. To determine the smallest eigenvalue, which is what we usually want, the process can be applied to the inverse matrix. A more elaborate algorithm that is harder to understand is the *QR method*, which in principle produces all the eigenvalues. The reader is referred to the literature [2] on algebraic algorithms for details.

3.4. ITERATIVE SOLUTION METHODS

General algebraic methods and methods such as that in the preceding section exist for finding the eigenvalues by direct calculation from the matrix representation. However, we clearly need a better method for very large matrices, and iterative methods are one such approach. To utilize an iterative method, we need equations to determine $E(i,j)$ and k. The equation for $E(i,j)$ can be derived from (3.9) as

$$E(i,j) = \frac{E(i+1,j) + E(i-1,j) + E(i,j+1) + E(i,j-1)}{\alpha} \tag{3-33}$$

if α is not zero. If the values of E were known exactly, (3-15) could be solved if α is not zero. If the values of E were known exactly, (3-15) could be solved for k^2 to give

$$k^2 = -\frac{\nabla^2 E}{E} \tag{3-34}$$

for any value of (i, j). A useful equation which, in a sense, is a weighted average can be derived by multiplication of (3-11) by E and integration over the entire region to give

$$\iint E\,\nabla^2 E\,dx\,dy + k^2 \iint E^2\,dx\,dy. \tag{3-35}$$

Solution for k^2 yields

$$k^2 = -\frac{\iint E\,\nabla^2 E\,dx\,dy}{\iint E^2\,dx\,dy}. \tag{3-36}$$

Use of the finite-difference approximation in this expression gives

$$k^2h^2 = -\frac{\sum_i \sum_j E(i,j)[E(i+1,j) + E(i-1,j) + E(i,j+1) + E(i,j-1) - 4E(i,j)]}{\sum_i \sum_j [E(i,j)]^2}. \tag{3-37}$$

Use of (3.16) simplifies this equation to

$$\alpha = \frac{\sum_i \sum_j E(i,j)[E(i+1,j) + E(i-1,j) + E(i,j+1) + E(i,j-1)]}{\sum_i \sum_j [E(i,j)]^2}. \tag{3-38}$$

Alternate use of (3.33) and (3.38) should converge to a solution for E and k. Which mode is converged to depends on the initial conditions. For example, with the initial condition that E is set to unity at all interior points, the convergence is usually to the lowest mode. Many initial conditions seem to converge to the lowest mode. To converge to other modes, one must select carefully the initial conditions, and this is probably not practical. A FORTRAN program for TM modes that uses unity for initial conditions follows.

```
      program wgtm
C     iterative solution for eigenvalue of TM
      mode for rectangular waveguide
      real f(130,130),Q(10,10)
      real R,tol,num,den,beta,ka,oldka
      integer M,i,j,n,nn,W,WW,C,Mx
      print*,'how many values of M?'
      read*,WW
      print*,'what is tolerance?'
      read*,tol
      R=0.7
      M=1
      do W=1,WW
         print*,'  iterations  alpha    ka'
         print*
         doj=0,2*M
           doi=0,4*M
             f(i,j)=0.
           end do
         end do
         do j=1,2*M-1
           do i=1,4*M-1
             f(i,j)=1.
           end do
         end do
         n=1
         nn=1
         ka=1
         oldka=0
         do while (abs(ka-oldka).GT.tol)
           oldka=ka
           do while (n.LE.nn)
```

```
        num=0
        den=0
        do j=1,2*M-1
            do i=1,4*M-1
                beta=f(i+1,j)+f(i-1,j)
                beta=beta+f(i,j+1)+f(i,j-1)
                num-num+f(i,j)*beta
                den=den+f(i,j)*f(i,j)
            end do
        end do
        alpha=num/den
        ka=4*M*sqrt(4-alpha)
        do j=1,2*M-1
            do i=1,4*M-1
                beta=f(i+1,j)+f(i-1,j)
                beta=beta+f(i,j+1)+f(i,j-1)
                f(i,j)=f(i,j)+R*(beta/alpha-
                f(i,j))
            end do
        end do
        n=n+1
      end do
      print*,(n-1),alpha,ka
      nn=2*nn
    end do
    Q(1,W)=ka
    C=1
    do i=2,W
      C=2*C
      Q(i,W)=(C*Q(i-1,W)-Q(i-1,W-1))/(C-1)
    end do
    do i=1,25
      print*
    end do
    print*,'      TM mode of rectangular
                  waveguide'
    print*,'    iterative finite-difference
                solution'
    print*,'     tolerance is',tol
```

```
      print*
      print*,'    M    ka    extrapolation'
      print*
      Mx=1
      do j=1,W
         print*,Mx,Q(1,j),Q(j,j)
         Mx=2*Mx
      end do
      M=2*M
   end do
   end
```

PROBLEMS

1. Use the finite-difference method to find an approximate value of the lowest normalized eigenvalue ka for the TM mode defined by

$$\nabla^2 E_z + k^2 E_z = 0,$$

with $E_z = 0$ on the boundary, for the mesh shown in the figure. Compare with the analytic solution.

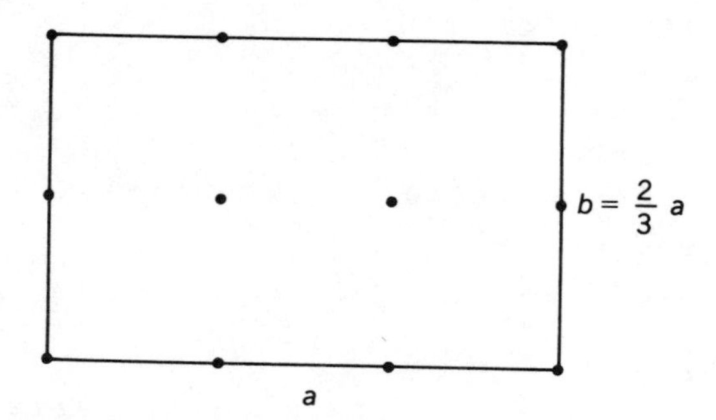

PROBLEM 3-1. Mesh for eigenvalue calculation.

2. Use the finite-difference method to find an approximate value of the lowest normalized eigenvalue ka for the waveguide TM

mode defined by

$$\nabla^2 E_z + k^2 E_z = 0,$$

with $E_z = 0$ on the boundary, for the mesh shown in the figure. Note that the spatial intervals h_x and h_y are not equal. Compare with the analytic solution.

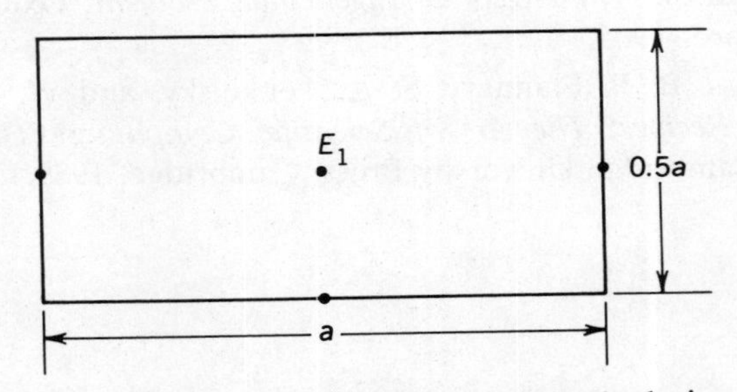

PROBLEM 3-2. Mesh for eigenvalue calculation.

3. Use the finite-difference method to find approximate values of the normalized eigenvalue ka for the nonrectangular waveguide

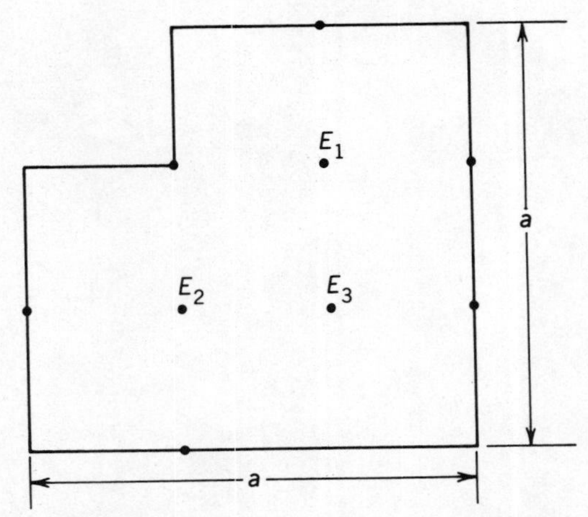

PROBLEM 3-3. Mesh for eigenvalue calculation.

TM mode defined by

$$\nabla^2 E_z + k^2 E_z = 0,$$

with $E_z = 0$ on the boundary, for the mesh shown in the figure.

REFERENCES

1. J. H. Wilkinson, *The Algebraic Eigenvalue Problem*, Oxford University Press, Oxford, 1965.
2. W. H. Press, B. P. Flannery, S. A. Teukolsky, and W. T. Vetterling, *Numerical Recipes: The Art of Scientific Computing (The FORTRAN Version)*, Cambridge University Press, Cambridge, 1989.

CHAPTER FOUR

Finite-Difference Time-Domain Method

4.1. WAVE EQUATION IN ONE SPATIAL DIMENSION

The finite-difference time-domain method is especially useful for the solution of wave equations. Let us begin with the simplest situation, that of one spatial dimension, where the wave equation is

$$\frac{\partial^2 E}{\partial t^2} = c^2 \frac{\partial^2 E}{\partial x^2}. \tag{4-1}$$

The general solution of this equation consists of two waves, one traveling in the positive x-direction, so

$$E(x,t) = E_1(x - ct) + E_2(x + ct). \tag{4-2}$$

The two functions on the right side of (4-2) are determined by the boundary and initial conditions. The boundary conditions require knowledge of

$$E(0,t) = b(t) \qquad \text{and} \qquad E(a,t) = c(t) \tag{4-3}$$

and the initial conditions are

$$E(x,0) = f(x) \qquad \text{and} \qquad \frac{\partial}{\partial t} E(x,0) = g(x). \tag{4-4}$$

If the finite-difference approach is used with the spatial variable x, and we take h as

$$h = \frac{a}{N}, \tag{4-5}$$

the central difference for the second derivative leads to

$$E(i+1,t) - 2E(i,t) + E(i-1,t) = \frac{h^2}{c^2}\frac{d^2E(i,t)}{dt^2}, \tag{4-6}$$

where the partial differential equation now has been replaced by a set of ordinary differential equations. There are a total of $N+1$ points, at two of which E is known from the boundary conditions, and $N-1$ locations at which E must be determined from the $N-1$ ordinary differential equations defined by (4-6). Although it is sometimes appropriate to solve these ordinary differential equations analytically, usually a numerical solution is more useful.

4.2. TIME QUANTIZATION

In addition to quantizing the spatial variables, the time variable can be quantized in the same way using a time interval δ. The field variable as a function of discrete indices we now denote by $E(i,n)$, and application of the central difference approximation to the derivative with respect to time leads to

$$\frac{E(i,n+1) - 2E(i,n) + E(i,n-1)}{\delta^2} = c^2\frac{E(i+1,n) - 2E(i,n) + E(i-1,n)}{h^2}. \tag{4-7}$$

Some rearrangement gives

$$E(i,n+1) - 2E(i,n) + E(i,n-1) = \rho^2[E(i+1,n) - 2F(i,n) + F(i-1,n)], \tag{4-8}$$

where

$$\rho = \frac{c\delta}{h}. \tag{4-9}$$

It is convenient to rearrange this equation for computation into the form

$$E(i, n+1) = (2 - 2\rho^2)E(i, n) - E(i, n-1) + \rho^2 S(i, n), \tag{4-10}$$

where

$$S(i, n) = E(i-1, n) + E(i+1, n). \tag{4-11}$$

E is determined at the endpoints from the given boundary conditions,

$$E(0, n) = b(n\delta) \quad \text{and} \quad E(N, n) = c(n\delta). \tag{4-12}$$

Once E is determined for $n = 0$ and for $n = 1$ from the initial conditions, later values can be calculated explicitly from (4-10) and (4-11). It can be shown that this numerical process is stable if

$$\rho \leq 1. \tag{4-13}$$

Now we wish to investigate the conditions under which the finite-difference approximations that we have used would be expected to be reasonably accurate. To investigate these questions, let us consider sinusoidal solutions. One set of continuous solutions has the form

$$E(x, t) = \sin\left(\frac{\omega}{c}x - \omega t\right). \tag{4-14}$$

The discrete solution can be described by

$$E(i, n) = \sin\left(\frac{\omega}{c}ih - n\omega\delta\right). \tag{4-15}$$

Let us investigate the circumstances under which the finite-difference approximation for the second derivative with respect to x is valid. From the continuous expression (4-14),

$$\frac{\partial^2 E}{\partial x^2} = -\left(\frac{\omega}{c}\right)^2 \sin\left(\frac{\omega}{c}x - \omega t\right), \tag{4-16}$$

and for the discrete solution (4-15), some trigonometric manipulation leads to

$$\frac{E(i+1,n) - 2E(i,n) + E(i-1,n)}{h^2} = \frac{2}{h^2}\left(\cos\frac{\omega h}{c} - 1\right)\sin\left(\frac{\omega x}{c} - \omega t\right). \tag{4-17}$$

The approximation for the second derivative clearly is valid if the approximation

$$\cos\frac{\omega h}{c} - 1 = -\frac{\omega^2 h^2}{2c^2} \tag{4-18}$$

is valid. This approximation is good to 1 percent if $\omega h/c$ is limited to approximately

$$\left(\frac{\omega h}{c}\right)_{\max} = 0.35. \tag{4-19}$$

In terms of the wavelength λ, this limit is approximately

$$\frac{h}{\lambda} \leq 0.055, \tag{4-20}$$

which means that we should have at least 18 points per wavelength, which seems reasonable. A similar calculation for the approximation to the time derivative leads to

$$(\omega\delta)_{\max} = 0.35. \tag{4-21}$$

Because

$$\omega\delta = \rho k h \tag{4-22}$$

and ρ must be less than 1 for stability, (4-21) is satisfied automatically when (4-19) is satisfied. The conditions just derived show when we would expect a reasonably accurate representation of the continuous solution.

4.3. INITIAL CONDITIONS

In addition to the differential equation, two boundary conditions valid for all time t and two initial conditions defined for all x are required to specify the solution. The two initial conditions are usually taken as the value of the function for time zero and the value of the first derivative with respect to time evaluated at time zero. Translation of the boundary conditions and the initial value of the function into conditions on the discrete function is given by (4-3) and (4-4). Use of the first derivative initial condition, however, needs a little more effort. We could use the first difference approximation

$$\frac{\partial E}{\partial t} = \frac{E(x,\delta) - E(x,0)}{\delta}, \tag{4-23}$$

but as we saw earlier, this approximation is accurate only to the first order in δ. A second-order approximation can be found from the Taylor series expansion

$$E(x,\delta) = E(x,0) + \delta \frac{\partial E}{\partial t}(x,0) + \frac{\delta^2}{2}\frac{\partial^2 E}{\partial t^2}(x,0) + \cdots . \tag{4-24}$$

Substitution from the wave equation gives

$$E(x,\delta) = E(x,0) + \delta \frac{\partial E}{\partial t}(x,0) + \frac{c^2\delta^2}{2}\frac{\partial^2 E}{\partial x^2}(x,0) + \cdots . \tag{4-25}$$

Introducing the finite-difference approximation for the derivative with respect to x yields

$$E(x,\delta) = E(x,0) + \delta\frac{\partial E}{\partial t}(x,0) + \frac{c^2\delta^2}{2h^2} \times [E(x-h,0) - 2E(x,0) + E(x+h,0)] + \cdots, \tag{4-26}$$

which can be rearranged to give

$$E(x,\delta) = (1-\rho^2)E(x,0) + \delta g(x) + \rho^2 S(x,0), \tag{4-27}$$

where

$$S(x,0) = E(x-h,0) + E(x+h,0). \tag{4-28}$$

4.4. WAVES IN TWO AND THREE SPATIAL DIMENSIONS

Although we have concentrated so far on the situation with one spatial dimension, the same equations with appropriate modifications hold for two and three spatial dimensions. The finite-difference approximations for the derivatives with respect to y and z follow the same pattern. The relation for one spatial dimension, (4-12), becomes

$$F(i,j,n+1) = (2-4\rho^2)E(i,j,n) - E(i,j,n-1) + \rho^2 S(i,j,n), \tag{4-29}$$

where

$$S(i,j,n) = E(i-1,j,n) + E(i+1,j,n) + E(i,j-1,n) + E(i,j+1,n) \tag{4-30}$$

and

$$\rho = \frac{c\delta}{h}. \tag{4-31}$$

The condition in two dimensions for stability is

$$\rho^2 \leq \tfrac{1}{2}, \tag{4-32}$$

which is approximately

$$\rho \leq 0.707. \tag{4-33}$$

The initial condition becomes

$$E(x, y, 0) = f(x, y), \tag{4-34}$$

and the equation for $E(x, y, \delta)$ following the same pattern as the one dimensional case becomes

$$E(x, y, \delta) = (1 - 2\rho^2)E(x, y, 0) + \delta g(x, y) + \frac{\rho^2}{2}S(x, y, 0). \tag{4-35}$$

As an example, solution for the TM waves in a rectangular waveguide leads to a program like that shown below (which does not include a provision for graphical output).

FORTRAN Program for Calculating TM Wave in Rectangular Waveguide Using Finite-Difference Time-Domain Method and for Calculating the Frequency Spectrum

```
      program wgtm
C     calculates two-dimensional TM wave and its
      frequency spectrum
C     for rectangular waveguide
      integer i,j,k,M,n,nt,nf
      real E(0:64,0:32),F(0:64,0:32)
      real G(0:64,0:32)
      real rho,z,R(0:3200)
      real alpha,d,ad,Re,Im,Mag(1600),pi,w,df
```

```
      pi=4.* atan(1.0)
      rho=sqrt(0.5)
C     input parameters
      print*,'what is M?'
      read *,M
      print*,'what is initial distribution param-
      eter alpha?'
      read*,alpha
      print*,'how many time points?'
      read*,nt
      print*,'how many frequency points?'
      read*,nf
      df=4.0*M/nt/rho
C     establish boundary conditions
      do i=0,4*M
         do j=0,2*M
            G(i,j)=0.
            F(i,j)=0.
            E(i,j)=0.
         end do
      end do
C     establish initial conditions
      do i=1,4*M-1
         do j=1,2*M-1
            d=sqrt((i-M)*(i-M)+(j-M)*(j-M))
            ad=alpha*d
            if (ad.LT.5) then
                G(i,j)=exp(-ad*ad)
            else
                G(i,j)=0.
            end if
         end do
      end do
      R(0)=G(3*M,M)
C     calculate values for n=1
      do i=1,4 *M-1
         do j=1,2 * M-1
            S=G(i-1,j)+G(i+1,j)+G(i,j-1)+G(i,j+1)
            z=(1-2*rho* ho)*G(i,j)
```

```
            z=z+rho*rho/2.* S
            F(i,j)=z
         end do
      end do
      R(1)=F(3*M,M)
C     print first two values of response
      do n=0,1
         print*,n,R(n)
      end do
C     calculate for remaining values of n
      do n=2,nt-1
         do i=1,4*M-1
            do j=1,2 *M-1
                S=F(i-1,j)+F(i+1,j)+F(i,j-1)+
                F(i,j+1)
                z=(2-4*rho*rho)* F(i,j)-G(i,j)
                z=z+rho*rho* S
                E(i,j)=z
            end do
         end do
         R(n)=E(3*M,M)
         print*,n,R(n)
         doi=1,4*M-1
            doj=1,2*M-1
                G(i,j)=F(i,j)
                F(i,j)=E(i,j)
            end do
         end do
      end do
C     calculate frequency spectrum
      do k=0,nf-1
         Re=0.
         Im=0.
         w=2.*pi/nt * k
         do n=0,nt-1
            Re=Re+R(n)*cos(w*n)
            Im=Im+R(n)*sin(w*n)
         end do
         Mag(k)=sqrt(Re*Re+Im*Im)
```

```
    print*,k,df*k,Mag(k)
end do
end
```

4.5. MAXWELL'S EQUATIONS

Thus far, we have dealt with the wave equation describing one scalar variable E, and thus we have dealt with an equation that involves second derivatives with respect to time. An alternative approach is to write two first-order equations. Although this can be done abstractly, for electromagnetic fields the curl equations of Maxwell introduce this approach with physical meaning for the variables involved. Since Yee's pioneering paper [1], a number of applications of this method have been made. As a simple one-dimensional example, if the fields are uniform with respect to x and y, the curl equations reduce to equations involving first derivatives with respect to z and t. One such field is described by

$$\frac{\partial}{\partial z}E_x = -\mu\frac{\partial}{\partial t}H_y \tag{4-36}$$

and

$$-\frac{\partial}{\partial z}H_y = \epsilon\frac{\partial}{\partial t}E_x. \tag{4-37}$$

Although with these continuous equations, we think of the two variables E_x and H_y as being defined at the same point in z and t, with the finite-difference equations this is not the case. In the first equation, for example, the derivative with respect to z is best thought of as being evaluated between the values of E_x, and thus the values of H_y are located in distance z between those values. The equation, however, locates the derivative of H_y with respect to time t there. The derivative with respect to time of H_y should be evaluated in time between the values of H_y. Similar considerations can be drawn from the finite-difference version of the second

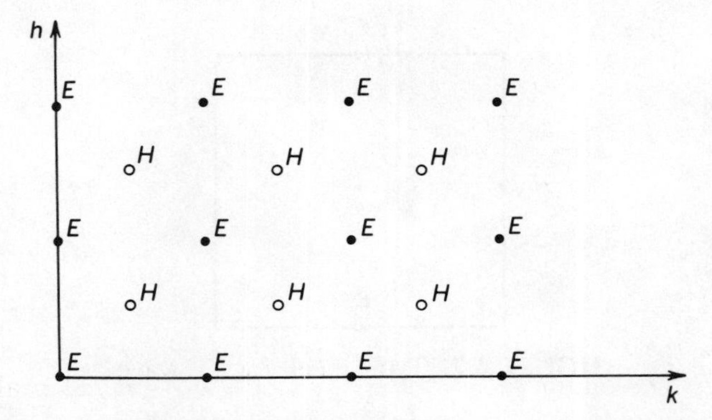

FIGURE 4-1. Mesh for Maxwell's equations in one spatial dimension.

equation. The situation is summarized by Fig. 4-1. The finite-difference approximations to (4-36) and (4-37) are

$$\frac{E(k+1,n)-E(k,n)}{h}$$
$$= -\mu \frac{H(k+0.5,n+0.5)-H(k+0.5,n-0.5)}{\delta} \tag{4-38}$$

and

$$-\frac{H(k+0.5,n+0.5)-H(k-0.5,n+0.5)}{h}$$
$$= \epsilon \frac{E(k,n+1)-E(k,n)}{\delta}. \tag{4-39}$$

The first equation corresponds to the continuous equation at $(k+0.5,n)$ and the second equation corresponds to $(k,n+0.5)$. Initial conditions can be imposed as values of E for $n=0$ and as values of H for $n=0.5$.

The situation in two and three spatial equations follows the same principles, except that we must decide where to position the points at which the field variables are defined. As first shown by Yee,

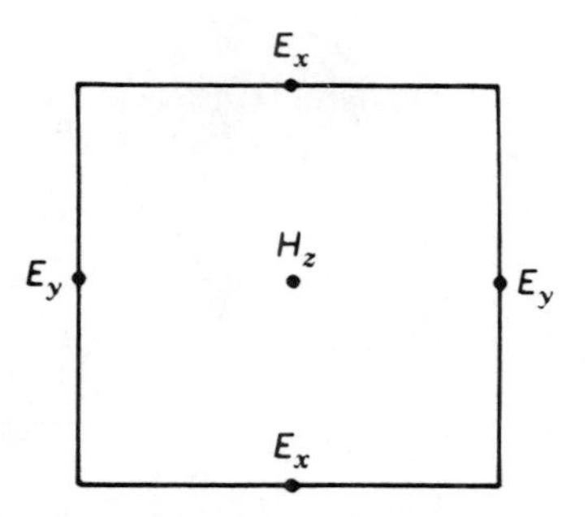

FIGURE 4-2. Typical cell for TE wave

these points can be so positioned that the finite-difference approximations are central-difference approximations, which result in second-order accuracy. If a two-dimensional region is divided into square cells, Fig. 4-2 shows one arrangement of these points in a typical cell for the TE situation which involves E_x, E_y, and H_z. Although the relative position of the points at which the field variables are defined is fixed, their positions in the cell and thus with respect to the boundaries of the region can take several forms. For example, although the typical cell for the TM situation could take the form of a dual of Fig. 4-2, as shown in Fig. 4-3, a more convenient arrangement is shown in Fig. 4-4. The latter arrangement results in E_z being on the boundary, which is appropriate for a metallic-wall boundary. Just as in the one-spatial dimension case,

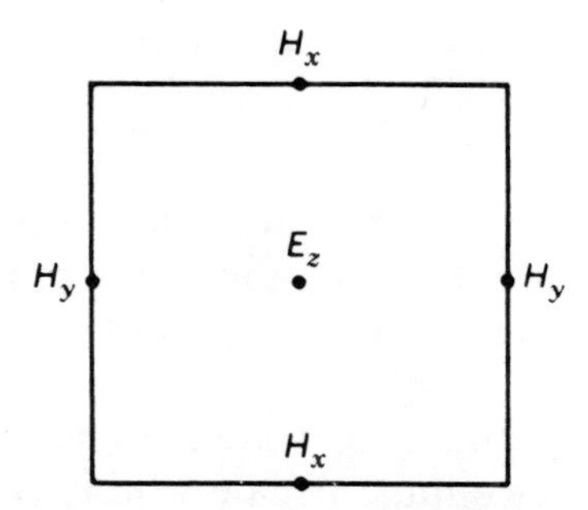

FIGURE 4-3. One version of the typical cell for TM wave.

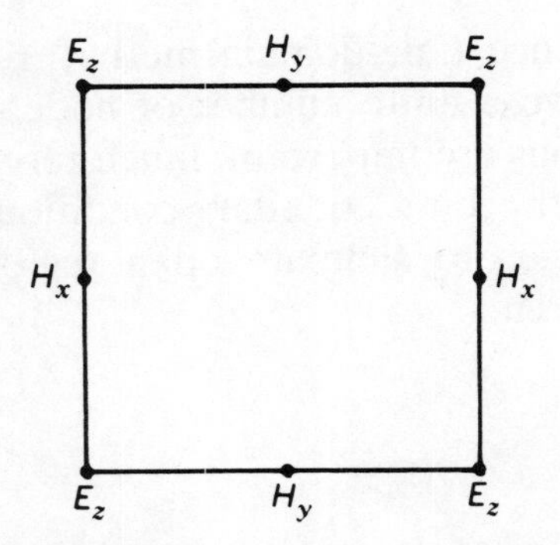

FIGURE 4-4. Another version of the typical cell for TM wave.

magnetic field values are evaluated at times intermediate to the times at which the electric field values are evaluated.

The situation in three spatial equations can be arranged as shown in Fig. 4-5. The three electric field components and three magnetic field components are positioned to satisfy the requirements for second-order approximations. In addition to their spatial location, magnetic field values again are evaluated at times intermediate to the times at which the electric field values are evaluated.

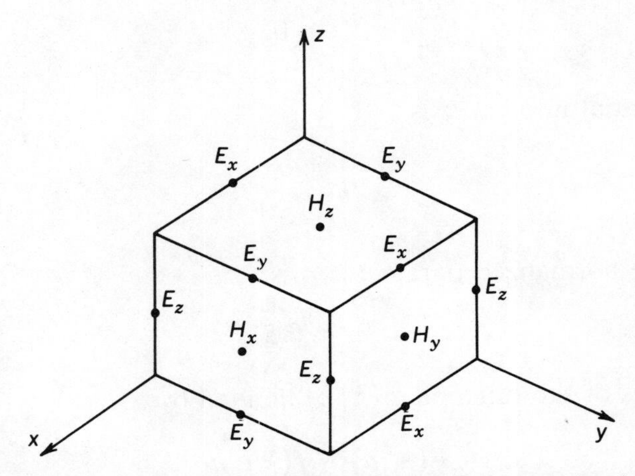

FIGURE 4-5. Spatial arrangement of field components in three spatial dimensions.

The finite-difference time-domain method requires a closed region in order to have a finite number of nodes. Because problems involving open regions are important, much research has performed to derive special "absorbing boundary conditions" to be applied at the region boundaries to simulate open regions. This is still an active area of research.

PROBLEM

1. The finite-difference time-domain method is to be applied to the determination of $f(x, t)$, where

$$\frac{\partial^2 f}{\partial x^2} = \frac{1}{c^2}\frac{\partial^2 f}{\partial t^2}$$

with the boundary conditions

$$f(0) = 0 \qquad \text{and} \qquad f(a) = 0$$

and the initial conditions

$$f(x,0) = \sin \pi x/a.$$

$$\frac{\partial f}{\partial t}(x,0) = 0.$$

The spatial interval h is

$$h = \frac{a}{4}$$

and the normalized parameter ρ is

$$\rho = 0.5.$$

In terms of the function $F(k, n)$ defined by

$$F(k,n) = f(kh, n\delta),$$

calculate $F(2,2)$.

COMPUTER PROJECT 4-1

The purpose of this project is to examine the nature of waves in two dimensions. Consider the TM electric field component $E_z(x, y, t)$, which satisfies the two-dimensional wave equation

$$\frac{\partial^2 E_z}{\partial x^2} + \frac{\partial^2 E_z}{\partial y^2} = \frac{1}{c^2}\frac{\partial^2 E_z}{\partial t^2},$$

in a square with sides L, with the initial conditions

$$E(x, y, t) = e^{-\alpha^2 R^2}, \qquad \text{where } R = \sqrt{(x - L/4)^2 + (y - L/2)^2}$$

and

$$\frac{\partial E}{\partial t}(x, y, 0) = 0.$$

The initial condition will propagate away from the vicinity of $(L/4, L/2)$ and the resulting "pulse" observed at $(L/2, L/2)$. With an appropriate value for α, the pulse will be sharp enough (well enough localized) to be recognizable. Select (by experimentation if necessary) a suitable value for α and evaluate the received wave $E(L/2, L/2, t)$ for a long enough period of time to see both the direct pulse and the first echo from the walls. Plot the response as a function of the normalized time ct/L. What can you say about the shape of the received pulse? What about the echoes?

REFERENCE

1. K. S. Yee, "Numerical solution of initial boundary value problems involving Maxwell's equations in isotropic media," *IEEE Transactions on Antennas and Propagation*, AP-14, pp. 302–307, 1966.

CHAPTER FIVE

Variational and Related Methods

5.1. STATIONARY CONDITIONS FOR FUNCTIONALS

The most common method for analyzing electromagnetic problems involves solving partial differential equations. Variational methods offer an alternative. A key concept in the variational calculus is that of a functional. This concept is a generalization of that of a function. Just as a function associates with each value of the independent variable(s) a numerical value, a functional associates with each function a numerical value. As a simple example, the length of curve defined by the function $f(x)$ can be expressed as

$$L[f(x)] = \int_a^b \sqrt{1 + \left(\frac{df}{dx}\right)^2}\, dx. \tag{5-1}$$

This is a special case of the functional

$$I[f] = \int_a^b F\left(x, f, \frac{df}{dx}\right) dx. \tag{5-2}$$

The next key concept is that certain functions make the functional achieve a maximum or a minimum value. For example, the true potential ϕ is known to minimize the electrostatic field energy.

The achievement of a maximum or minimum value, together with the equivalent of the saddle point (a function for which first-order variations of the functional are zero and yet the functional is neither maximum nor minimum), is referred to as *stationarity*.

One method for determining the function that makes the functional stationary is to substitute for f the function $f + \alpha g$, where g is a function that is arbitrary except that it satisfies homogeneous boundary conditions and α is a parameter. Then f makes the functional stationary if and only if

$$\frac{d}{d\alpha} I[f + \alpha g] = 0 \qquad \text{for } \alpha = 0. \tag{5-3}$$

As a simple one-dimensional example, to determine the minimum of

$$I[f] = \int_0^1 \left(\frac{df}{dx}\right)^2 dx, \qquad \text{with } f(0) = 0 \text{ and } f(1) = 1, \tag{5-4}$$

we replace f by $f + \alpha g$, where $g(0) = 0$ and $g(1) = 0$. Then the integral can be written as

$$I[f + \alpha g] + I[f] + 2\alpha\left[\int_0^1 \left(\frac{df}{dx}\right)\left(\frac{dg}{dx}\right) dx\right] + \alpha^2 I[g]. \tag{5-5}$$

Use of (5-3) gives

$$\int_0^1 \left(\frac{df}{dx}\right)\left(\frac{dg}{dx}\right) dx = 0. \tag{5-6}$$

Integration by parts results in

$$\left[\frac{df}{dx} g\right]_0^1 - \int_0^1 \left(\frac{d^2 f}{dx^2}\right) g\, dx = 0. \tag{5-7}$$

The bracketed expression is zero because of the boundary conditions on g, and thus

$$\int_0^1 g\left(\frac{d^2f}{dx^2}\right)g\,dx = 0. \tag{5-8}$$

For this integral to be zero for arbitrary functions g,

$$\frac{d^2f}{dx^2} = 0. \tag{5-9}$$

The more general functional in (5-2) can be analyzed in a similar way and leads to the differential equation

$$F_{f'}f'' + F_f f' + F_x = 0, \tag{5-10}$$

known as the *Euler–Lagrange equation*. In a similar manner, many functionals that are in the form of integrals are made stationary by functions that satisfy differential equations and this same approach can be used to derive the differential equation. For example, the functional

$$I[f] = \int_0^1 \left[\left(\frac{df}{dx}\right)^2 + f^2\right]dx, \qquad \text{with } f(0) = 0 \quad \text{and} \quad f(1) = 1, \tag{5-11}$$

becomes

$$I[f+\alpha g] = I[f] + 2\alpha\left[\int_0^1 \left(\frac{df}{dx}\right)\left(\frac{dg}{dx}\right)dx + \int_0^1 fg\,dx\right] + \alpha^2 I[g]. \tag{5-12}$$

The function f then can be shown to satisfy

$$\frac{d^2f}{dx^2} - f = 0. \tag{5-13}$$

5.2. EXAMPLE OF ELECTROSTATIC FIELD ENERGY

An important example from electrostatics is the electrical field energy,

$$W[\phi(x, y, z)] = \frac{1}{2}\int\int\int \epsilon\left[\left(\frac{\partial\phi}{\partial x}\right)^2 + \left(\frac{\partial\phi}{\partial y}\right)^2 + \left(\frac{\partial\phi}{\partial z}\right)^2\right] dx\, dy\, dz, \tag{5-14}$$

where the permittivity $\epsilon = \epsilon(x, y, z)$ in general is a function of position. Use of the same method as in Section 5.1 results in

$$W[\phi + \alpha\psi] = I[\phi] + 2\alpha\int\int\int \epsilon\left[\left(\frac{\partial\phi}{\partial x}\right)\left(\frac{\partial\psi}{\partial x}\right) + \left(\frac{\partial\phi}{\partial y}\right)\left(\frac{\partial\psi}{\partial y}\right) + \left(\frac{\partial\phi}{\partial z}\right)\left(\frac{\partial\psi}{\partial z}\right)\right] dx\, dy\, dz + \alpha^2 I[\psi]. \tag{5-15}$$

Use of (5-3) again gives

$$\int\int\int \epsilon\left[\left(\frac{\partial\phi}{\partial x}\right)\left(\frac{\partial\psi}{\partial x}\right) + \left(\frac{\partial\phi}{\partial y}\right)\left(\frac{\partial\psi}{\partial y}\right) + \left(\frac{\partial\phi}{\partial z}\right)\left(\frac{\partial\psi}{\partial z}\right)\right] dx\, dy\, dz = 0. \tag{5-16}$$

This integral can be transformed by use of one of Green's theorems (essentially integration by parts) to

$$\int\int\int\left[\frac{\partial}{\partial x}\left(\epsilon\frac{\partial\phi}{\partial x}\right) + \frac{\partial}{\partial y}\left(\epsilon\frac{\partial\phi}{\partial y}\right) + \frac{\partial}{\partial z}\left(\epsilon\frac{\partial\phi}{\partial z}\right)\right]\psi\, dx\, dy\, dz = 0, \tag{5-17}$$

and thus

$$\left[\frac{\partial}{\partial x}\left(\epsilon\frac{\partial\phi}{\partial x}\right) + \frac{\partial}{\partial y}\left(\epsilon\frac{\partial\phi}{\partial y}\right) + \frac{\partial}{\partial z}\left(\epsilon\frac{\partial\phi}{\partial z}\right)\right] = 0. \tag{5-18}$$

or in vector notation,

$$\nabla \cdot (\epsilon \nabla \phi) = 0, \tag{5-19}$$

which is of course the generalized form of Laplace's equation that can be derived from

$$\nabla \cdot \mathbf{D} = 0. \tag{5-20}$$

5.3. VARIATIONAL EXPRESSION FOR EIGENVALUES

The preceding examples of functionals involve a single integral. A functional for eigenvalues involves the ratio of two integrals. Consider first the one-dimensional eigenvalue equation

$$\frac{d^2 f}{dx^2} + k^2 f = 0 \qquad \text{with } f(0) = 0 \quad \text{and} \quad f(1) = 0. \tag{5-21}$$

Multiplication by $f(x)$ and integration results in

$$\int_0^1 f \frac{d^2 f}{dx^2}\, dx + k^2 \int_0^1 f^2\, dx = 0, \tag{5-22}$$

from which

$$k^2 = -\frac{\displaystyle\int_0^1 f \frac{d^2 f}{dx^2}\, dx}{\displaystyle\int_0^1 f^2\, dx}. \tag{5-23}$$

With integration by parts of the numerator, this can also be written as

$$k^2 = \frac{\displaystyle\int_0^1 \left(\frac{df}{dx}\right)^2 dx}{\displaystyle\int_0^1 f^2\, dx}. \tag{5-24}$$

Expressions (5-23) and (5-24) are functionals that depend on the function $f(x)$ and are variational expressions. The functional is minimized by the eigenfunction and the minimum value is the eigenvalue. Starting with (5-23) we consider as before

$$k^2(f+\alpha g) = -\frac{\int_0^1 (f+\alpha g)\left(\frac{d^2 f}{dx^2} + \alpha \frac{d^2 g}{dx^2}\right)^2 dx}{\int_0^1 (f+\alpha g)^2 \, dx}. \tag{5-25}$$

If we set

$$\frac{\partial}{\partial \alpha} k^2(f+\alpha g)|_{\alpha=0} = 0, \tag{5-26}$$

the resulting equation can be simplified to

$$\frac{d^2 f}{dx^2}\int_0^1 f^2 \, dx = f\int_0^1 f\frac{d^2 f}{dx^2}\, dx, \tag{5-27}$$

which can be written in the form

$$\frac{d^2 f}{dx^2} - f\frac{\int_0^1 f\frac{d^2 f}{dx^2}\, dx}{\int_0^1 f^2 \, dx} = 0, \tag{5-28}$$

which with the notation of (5-23) is the same as (5-21). The minimization is of course only a local minimum and there are an infinite number of minima, one for each eigenvalue.

In two dimensions, (5-23) and (5-24) generalize to expressions of the form

$$k^2 = -\frac{\iint f\nabla^2 f \, dx\, dy}{\iint f^2 \, dx\, dy} \tag{5-29}$$

and

$$k^2 = \frac{\iint (\nabla f) \cdot (\nabla f)\, dx\, dy}{\iint f^2\, dx\, dy}. \tag{5-30}$$

which correspond to the differential equation

$$(\nabla^2 + k^2) f(x, y) = 0. \tag{5-31}$$

In three dimensions, the generalizations are

$$k^2 = \frac{\iiint f \nabla^2 f\, dx\, dy\, dz}{\iiint f^2\, dx\, dy\, dz} \tag{5-32}$$

and

$$k^2 = \frac{\iiint (\nabla f) \cdot (\nabla f)\, dx\, dy\, dz}{\iint f^2\, dx\, dy\, dz}. \tag{5-33}$$

which correspond to the differential equation

$$(\nabla^2 + k^2) f(x, y, z) = 0. \tag{5-34}$$

5.4. RAYLEIGH – RITZ METHOD

The *Rayleigh–Ritz method* is a general method to obtain an approximation to the function that makes a functional stationary. This method is based on pioneering work by Lord Rayleigh in 1870 and improvements by Ritz in 1909. The function is approximated by a linear combination of known functions and the solution consists of determining the parameters in the combination. This determination involves substitution of the linear combination into the functional and then differentiation with respect to each parameter.

Consider again the example

$$I[f] = \int_0^1 \left[\left(\frac{df}{dx} \right)^2 + f^2 \right] dx, \qquad \text{with } f(0) = 0 \text{ and } f(1) = 1, \tag{5-35}$$

A function that satisfies the boundary conditions but does not minimize the functional is

$$f_0(x) = x. \tag{5-36}$$

A set of functions that can be used to improve the approximation is described by

$$f_j(x) = x - x^{j+1}. \tag{5-37}$$

These functions satisfy the homogeneous boundary values

$$f_j(0) = 0 \qquad \text{and} \qquad f_j(1) = 0, \tag{5-38}$$

and hence a linear combination of these functions can be added to the right side of (5-36) without changing the boundary values. The resulting approximation is

$$f(x) = x + \alpha_1(x - x^2) + \alpha_2(x - x^3) + \cdots + \alpha_n(x - x^{N+1}) \tag{5-39}$$

or

$$f(x) = x + \sum_{j=1}^{N} \alpha_j(x - x^{j+1}). \tag{5-40}$$

A generalization of this expression is

$$f(x) = f_0(x) + \sum_{n=1}^{N} \alpha_n f_n(x), \tag{5-41}$$

where f_0 satisfies the nonhomogeneous boundary conditions, the f_n for $1 \le n \le N$ satisfy homogeneous boundary conditions, and the α_n are parameters to be determined. If the boundary conditions on

f are zero, as frequently is the case with eigenvalue problems, then f_0 is zero. The sum from (5-41) is substituted into the functional and the derivative of the resulting expression with respect to each of the parameters is set equal to zero, which yields a set of N equations. Solution of this set of equations yields the parameters and thus the approximate solution for the function that makes the functional stationary. Substitution into (5-35) yields

$$I[f] = \int_0^1 \left[\left(\frac{df_0}{dx} \right)^2 + f_0^2 \right] dx + 2 \sum_{j=1}^{N} \alpha_j \int_0^1 \left[\left(\frac{df_0}{dx} \right) \left(\frac{df_j}{dx} \right) + f_0(x) f_j(x) \right] dx + \sum_{j=1}^{N} \sum_{k=1}^{N} \alpha_j \alpha_k \int_0^1 \left[\left(\frac{df_j}{dx} \right) \left(\frac{df_k}{dx} \right) + f_j(x) f_k(x) \right] dx. \tag{5-42}$$

Differentiation with respect to each α_j gives

$$\sum_{k=1}^{N} \alpha_k \int_0^1 \left[\left(\frac{df_j}{dx} \right) \left(\frac{df_k}{dx} \right) + f_j(x) f_k(x) \right] dx = - \int_0^1 \left[\left(\frac{df_0}{dx} \right) \left(\frac{df_j}{dx} \right) + f_0(x) f_j(x) \right] dx. \tag{5-43}$$

This is a set of linear algebraic equations from which the parameters α_k can be determined.

PROBLEMS

1. Use the variational method on (5-1) to show that the shortest distance between two points in a plane is a straight line.

2. Use variational methods to show that the functional

$$I[f] = \int_0^1 \left[\left(\frac{df}{dx} \right)^2 - f^2 \right] dx,$$

with boundary conditions $f(0) = 0$ and $f(1) = 1$,

is made stationary by the solution of

$$\frac{d^2f}{dx^2} + f = 0 \qquad \text{(with the same boundary conditions).}$$

Note that this is not a minimum or maximum. Find the function $f(x)$ by solution of the differential equation and evaluate the resulting value of the functional I. Evaluate the functional for the trial function

$$f_1(x) = x$$

and compare the result with the previous value.

3. The functional I, defined by

$$I[f] = \int_0^1 \left(\frac{df}{dx}\right)^2 dx$$

with the boundary conditions

$$f(0) = 0 \qquad \text{and} \qquad f(1) = 1$$

is made stationary by the function

$$f(x) = x.$$

Express as a polynomial in α the functional for the function

$$f(x) = x + \alpha(x - x^2).$$

Sketch this polynomial. What conclusion you can draw from this sketch?

4. For the eigenvalue equation (5-21) and the corresponding functional (5-24), consider the trial function

$$f(x) = a + bx + cx^2.$$

Choose the coefficients a, b, and c so that the trial function satisfies the boundary conditions. Then evaluate the functional given by (5-24). Solve the differential equation (5-21) and determine the true value of k^2. What two conclusions would you draw from a comparison of these two values for k^2?

REFERENCE

1. R. Weinstock, *Calculus of Variations* (Reprint Edition), Dover Publications, New York, 1974.

CHAPTER SIX

Finite-Element Method

6.1. BASIC CONCEPT OF FINITE ELEMENTS

The finite-element method developed initially in mechanical and civil engineering and most of the applications and, accordingly, books and other publications have involved structures. The method is of increasing importance in other fields, including electromagnetics. Several books should be of use to those interested in electromagnetics [1–4].

The basic concept of the finite-element method is that although the behavior of a function may be complex when viewed over a large region, a simple approximation may suffice for a small subregion. The total region is divided into a number of nonoverlapping subregions called *finite elements*. In two dimensions we usually use polygons, and the simplest polygons are triangles and squares. Figure 6-1 shows a region divided into squares and Fig. 6-2 shows the same region divided into isoceles right triangles. Sometimes, as illustrated in Fig. 6-3, a combination of triangles and squares is useful. One of the advantages of using triangles is that a fairly arbitrary region can be more easily approximately covered by a set of triangles, as shown in Fig. 6-4. Regardless of the shape of the elements, the field is approximated by a different expression over each element, but where the edges of adjoining elements overlap, the field representations must agree to maintain continuity of the field. The equations to be solved are usually stated in terms not of the field variables but in terms of an integral-type functional such as

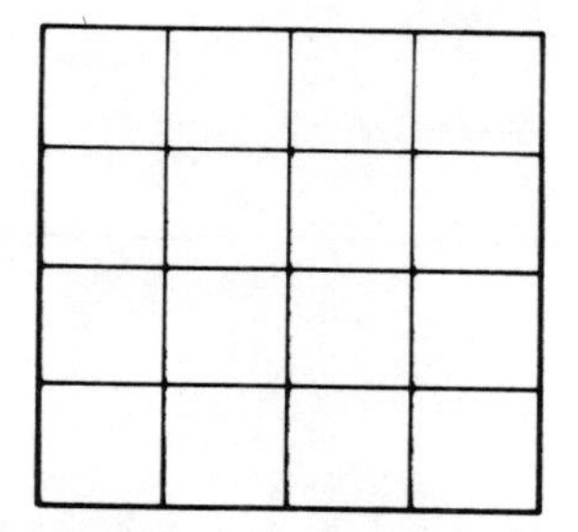

FIGURE 6-1. Division of a region into square elements.

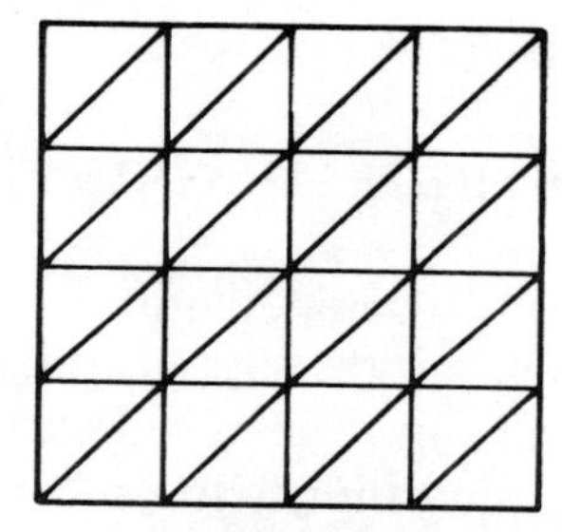

FIGURE 6-2. Division of a region into right isoceles triangles.

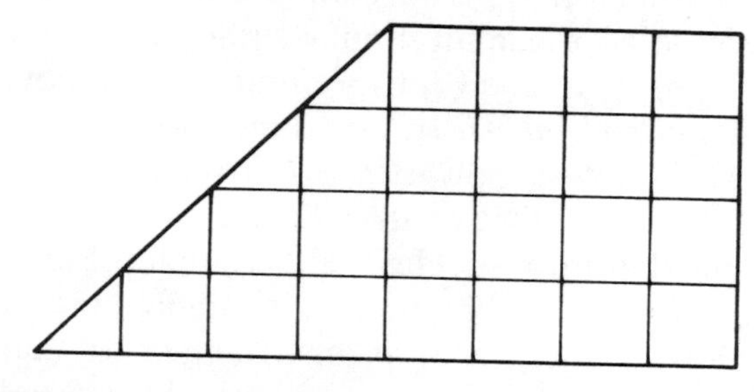

FIGURE 6-3. Division of a region into triangles and squares.

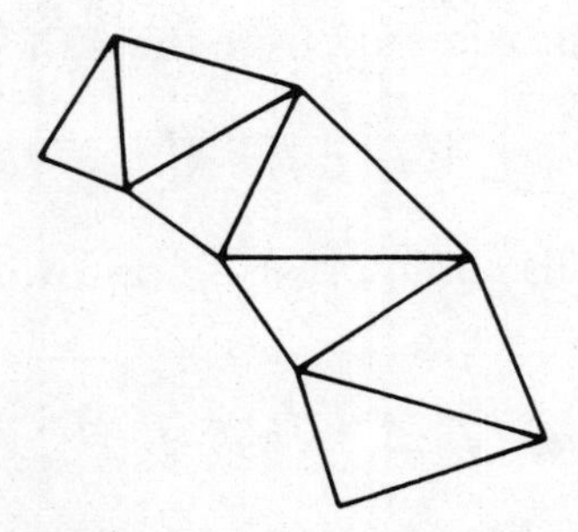

FIGURE 6-4. Division of an irregular region into triangles.

energy. The functional is chosen such that the field solution makes the functional stationary. The total functional is the sum of the integral over each element.

6.2. FINITE ELEMENTS IN ONE DIMENSION

Consider the problem in one dimension of minimizing the functional

$$I[f] = \int_0^1 \left[\frac{df}{dx}\right]^2 dx \tag{6-1}$$

with the boundary conditions

$$f(0) = 0 \tag{6-2}$$

and

$$f(1) = 1. \tag{6-3}$$

In a manner similar to that of Chapter 2, divide the interval (0, 1) into four subintervals and use the notation

$$f_k = f(0.25k). \tag{6-4}$$

If the function is approximated over the kth subinterval by

$$f(x) = f_{k-1} + 4(f_k - f_{k-1})[x - 0.25(k-1)], \tag{6-5}$$

the derivative over this subinterval is approximately

$$\frac{df}{dx} = 4(f_k - f_{k-1}). \tag{6-6}$$

Evaluation of the integrals over each of the subintervals yields

$$I[f] = 4(f_1)^2 + 4(f_2 - f_1)^2 + 4(f_3 - f_2)^2 + 4(1 - f_3)^2. \tag{6-7}$$

To find the functional values that minimize this expression, differentiation with respect to each of these values gives

$$\frac{dI}{dx} = 8(f_1) + 8(f_1 - f_2), \tag{6-8}$$

$$\frac{dI}{dx} = 8(f_2 - f_1) + 8(f_2 - f_3), \tag{6-9}$$

and

$$\frac{dI}{dx} = 8(f_3 - f_2) + 8(f_3 - 1). \tag{6-10}$$

Solution of these equations gives

$$f_1 = 0.25, \tag{6-11}$$

$$f_2 = 0.5, \tag{6-12}$$

and

$$f_3 = 0.75. \tag{6-13}$$

Substitution of these values into (6-7) gives the value of 1 for $I[f]$. Two observations can be made about this solution. First the equations derived by minimizing the approximate expression for the functional can be seen to be the same equations that result from application of the finite-difference method to the differential equation whose solution minimizes the functional. Second and probably not very important, the functional values and the minimum value of the functional derived by the finite-element method happen to be exact for this simple example.

Now consider the slightly more complex example

$$I[f] = \int_0^1 \left[\left(\frac{df}{dx} \right)^2 + (f)^2 \right] dx. \tag{6-14}$$

Let us use the same approximation as before with four elements (subintervals). To the previously derived value for the first functional, we must add the result of integrating the second term, which can be evaluated as

$$\int_0^1 f^2\, dx = \tfrac{1}{12}(2f_1^2 + f_1 f_2 + 2f_2^2 + f_2 f_3 + 2f_3^2 + f_3 + 1). \tag{6-15}$$

The total functional then is

$$\begin{aligned} I[f] &= 8.17f_1^2 - 7.92f_1 f_2 + 8.17f_2^2 - 7.92f_2 f_3 \\ &\quad + 8.17f_3^2 - 7.92f_3 + 4.08. \end{aligned} \tag{6-16}$$

Setting the derivatives of the functional with respect to f_1, f_2, and f_3 equal to zero leads to the equations

$$\begin{aligned} 16.34f_1 - 7.92f_2 &= 0 \\ -7.92f_1 + 16.34f_2 - 7.92f_3 &= 0 \\ -7.92f_2 + 16.34f_3 &= 7.92, \end{aligned} \tag{6-17}$$

which have the solutions

$$\begin{aligned} f_1 &= 0.21 \\ f_2 &= 0.44 \\ f_3 &= 0.70, \end{aligned} \tag{6-18}$$

which agree well with the analytic solution.

6.3. LINEAR INTERPOLATION FOR ISOCELES RIGHT TRIANGLES

In two dimensions the simplest element is the isoceles right triangle. Consider the typical such triangle shown in Fig. 6-5. For convenience while analyzing this triangle, the origin of the coordinates is taken at one vertex. The simplest interpolation to use is the first-order formula

$$\phi(x, y) = a + bx + cy. \tag{6-19}$$

There are three unknown parameters (a, b, and c), so it is natural to specify the interpolation in terms of the value of the function at each of the three vertices of the triangle. Solution for the parameters results in

$$\phi(x, y) = \phi_0 + \frac{\phi_1 - \phi_0}{h}x + \frac{\phi_2 - \phi_0}{h}y, \tag{6-20}$$

which can be written in the form

$$\phi(x, y) = \left(1 - \frac{x}{h} - \frac{y}{h}\right)\phi_0 + \left(\frac{x}{h}\right)\phi_1 + \left(\frac{y}{h}\right)\phi_2, \tag{6-21}$$

usually referred to as a *nodal expansion*. When the field quantity being sought is the electrostatic potential, the electric field energy is a natural functional to use because the potential is known to minimize this energy. This energy in two dimensions for uniform

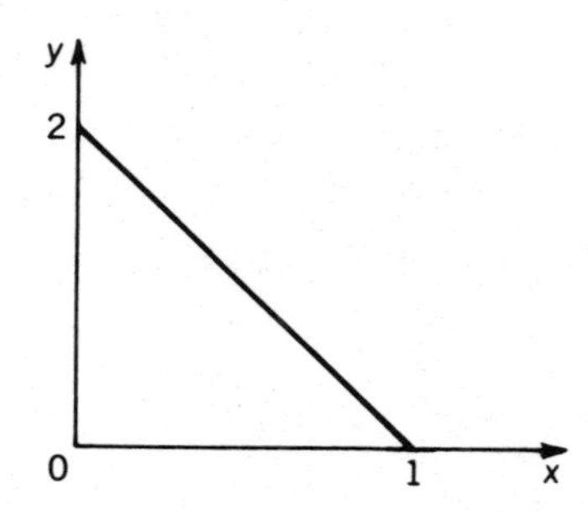

FIGURE 6-5. A typical right isoceles triangle.

permittivity ϵ is

$$W = \frac{\epsilon}{2} \iint (\nabla\phi) \cdot (\nabla\phi) \, dx \, dy, \tag{6-22}$$

which is

$$W = \frac{\epsilon}{2} \iint \left[\left(\frac{\partial \phi}{\partial x} \right)^2 + \left(\frac{\partial \phi}{\partial y} \right)^2 \right] dx \, dy. \tag{6-23}$$

The energy for the isoceles right triangle is found by substitution of (6-20) into (6.23), which yields

$$W = \frac{\epsilon}{4} \left[(\phi_0 - \phi_1)^2 + (\phi_0 - \phi_2)^2 \right]. \tag{6-24}$$

This expression clearly does not depend on the orientation of the triangle and applies to other orientations. If the total region is divided into a number of right isoceles triangles, the energy for each element is proportional to the sum of squares of the differences of the vertex values, using the general form of (6-24). The total energy then is found by addition of the energy contributions from each element.

The true potential minimizes the total energy, and therefore the approximate solution is determined by differentiation with respect to the parameters, which in this case are the unconstrained values

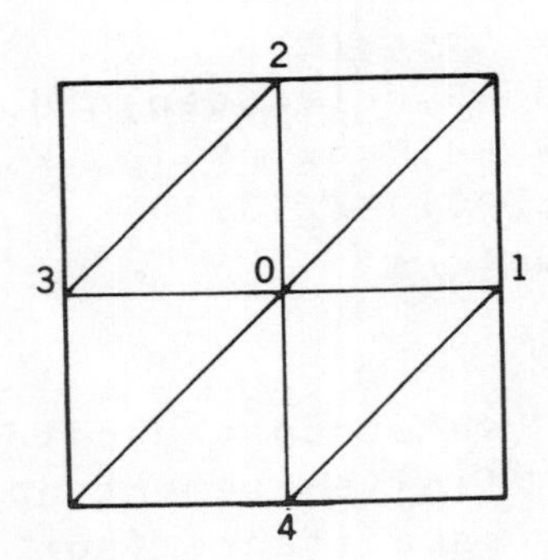

FIGURE 6-6. A typical interior node with triangular elements.

of the potential at the triangle vertices. Note that the boundary conditions constrain values at the nodes that fall on the boundary. The energy is the sum of a number of terms, some of which involve $\phi(i, j)$ and some of which do not. For a typical interior point (x, y) as illustrated in Fig. 6-6,

$$\text{energy terms that involve } \phi_0 = \frac{\epsilon}{2}[\phi_0 - \phi_1]^2 + \frac{\epsilon}{2}[\phi_0 - \phi_2]^2 + \frac{\epsilon}{2}[\phi_0 - \phi_3]^2 + \frac{\epsilon}{2}[\phi_0 - \phi_4]^2. \tag{6-25}$$

Setting the derivative with respect to ϕ_0 equal to zero yields

$$\phi_0 = \frac{\phi_1 + \phi_2 + \phi_3 + \phi_4}{4}, \tag{6-26}$$

which is the same equation derived earlier with the finite-difference method. The following FORTRAN program uses the finite-element method with isoceles right triangles to determine the shielded parallel-plate transmission line capacitance problem considered in Chapter 2.

```
      program shcapfe
c     capacitance of shielded capacitor
c     calculated with finite elements
c     line integral of E field
c     solution by iteration
      real add,bdd,R,delta,cap,old,z
      integer C,W,WW,M,nx,ny,i,j,k,kk
      real V(200,200),ZZ(10,10)
c     input parameters
      print*
      print*
      print*,'     shielded capacitor'
      print*,'     finite-element approximation'
      print*,'     capacitance from line integral'
      print*,'     solution by iteration'
```

```
      print*
      print*,'    what is a/d?'
      read*, add
      print*
      print*,'what is b/d?'
      read*,bdd
      print*
      print*,'how many values of M?'
      read*,WW
      print*
      print *,'   enter tolerance'
      read*,tol
      print*
      print *
      R=1.4
C     print parameters
      print*,'    shielded capacitor'
      print*,'    a/d is',add
      print*,'    b/d is',bdd
      print*,'    tolerance is',tol
      print*,''
      print*,'    M  original extrapolations'
      print*,''
C     start calculation
      M=1
      W=1
      do while (W.LE.WW)
      nx=add*M
      ny=bdd*M
C     set initial and boundary conditions
      do j=0,ny
         do i=0,nx
            V(i,j)=0
         end do
      end do
      do j=1,ny-1
         do i=0,nx-1
            V(i,j)=1
         end do
```

```
      end do
      cap=1
      old=0
      k=1
      kk=1
      do while (abs(old-cap).GT.tol)
         old=cap
         do while (k.LE.kk)
            do j=1,ny-1
               do i=0,nx-1
                 if (j.EQ.M) then
                     if (i.GT.M) then
                        z=V(i-1,j)+V(i+1,j)
                        z=z+V(i,j-1)+V(i,j+1)
                        delta=z/4-V(i,j)
                        V(i,j)=V(i,j)+R*delta
                     end if
                 else
                     if (i.EQ.0) then
                        z=2*V(i+1,j)+V(i,j-1)
                        z=z+V(i,j+1)
                        delta=z/4-V(i,j)
                        V(i,j)=V(i,j)+R*delta
                     else
                        z=V(i-1,j)+V(i+1,j)
                        z=z+V(i,j-1)+V(i,j+1)
                        delta=z/4-V(i,j)
                        V(i,j)=V(i,j)+R*delta
                     end if
                 end if
               end do
            end do
            k=k+1
         end do
C        start calculation of capacitance
         z=0
         do j=0,ny-1
            do i=0,nx-1
               z=z+(V(i+1,j)-V(i,j))*
```

```
                  (V(i+1,j)-V(i,j))
                  z=z+(V(i+1,j+1)-V(i,j+1))*
                  (V(i+1,j+1)-V(i,j+1))
                  z=z+(V(i,j+1)-V(i,j))*
                  (V(i,j+1)-V(i,j))
                  z=z+(V(i+1,j+1)-V(i+1,j))*
                  (V(i+1,j+1)-V(i+1,j))
               end do
            end do
            cap=8.854187*z/2.
            kk=2*kk
         end do
         ZZ(W,1)=cap
C        extrapolation calculation
         C=1
         do j=2,W
            C=2*C
            ZZ(W,j)=(C*ZZ(W,j-1)-ZZ(W-1,j-1))/(C-1)
         end do
C        print results
         print*,M,ZZ(W,1),ZZ(W,W)
         W=W+1
         M=2*M
      end do
      end
```

An interesting situation arises at boundary points for which the boundary condition is that the normal derivative is zero. For a point with unknown value on a boundary as illustrated in Fig. 6-7,

energy terms that involve ϕ_0

$$= \frac{\epsilon}{4}(\phi_0 - \phi_1)^2 + \frac{\epsilon}{4}(\phi_0 - \phi_3)^2 + \frac{\epsilon}{2}(\phi_0 - \phi_4)^2. \quad (6\text{-}27)$$

Setting the derivative with respect to $\phi(x, y)$ equal to zero yields

$$\phi_0 = \frac{\phi_1 + \phi_3 + 2\phi_4}{4}. \quad (6\text{-}28)$$

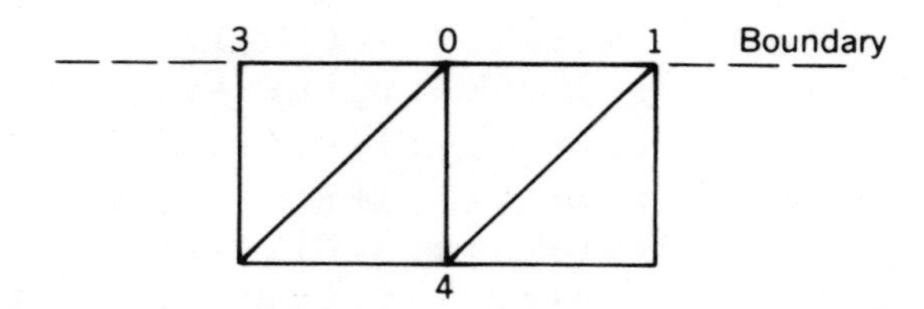

FIGURE 6-7. A node with unknown value on a boundary.

Note that although the boundary condition that the normal derivative is zero was not used, the result is the same as the finite-difference result, where we used that boundary condition explicitly.

6.4. SQUARE ELEMENTS

Consider now the square element shown in Fig. 6-8. One simple approach to determining an approximation to the energy of this element is to divide the square into two triangles, as indicated in Fig. 6-9. Addition of the energy for the two subelements gives

$$W = \frac{\epsilon}{4}\left[(\phi_0 - \phi_1)^2 + (\phi_3 - \phi_2)^2 + (\phi_0 - \phi_3)^2 + (\phi_1 - \phi_2)^2\right]. \tag{6-29}$$

This approach, although workable, does not correspond to a smooth

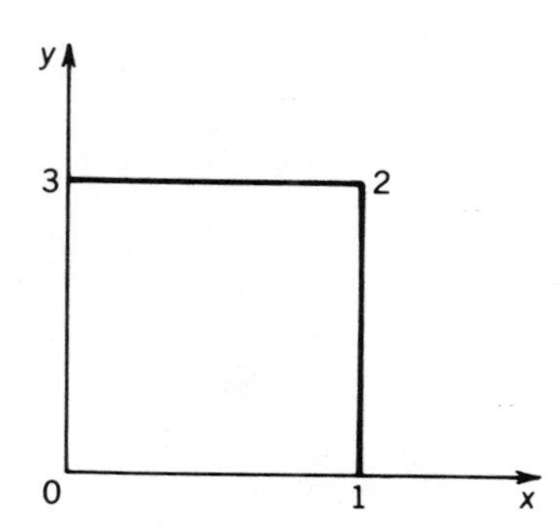

FIGURE 6-8. A typical square element.

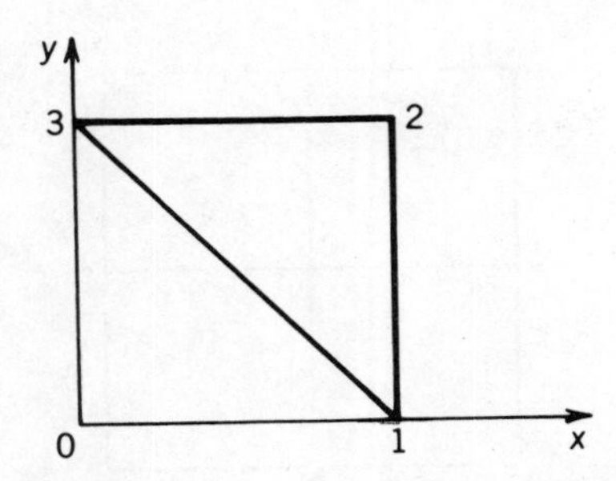

FIGURE 6-9. A square element divided into triangles.

interpolation over the square. Such an interpolation that uses the four nodal values should have four coefficients. One such approach is the bilinear interpolation

$$\phi(x, y) = a + bx + cy + dxy. \tag{6-30}$$

Evaluation of the four coefficients and rearranging as a nodal expansion gives

$$\begin{aligned}\phi(x, y) = {} & \left(1 - \frac{x}{h}\right)\left(1 - \frac{y}{h}\right)\phi_0 + \left(\frac{x}{h}\right)\left(1 - \frac{y}{h}\right)\phi_1 \\ & + \left(\frac{x}{h}\right)\left(\frac{y}{h}\right)\phi_2 + \left(1 - \frac{x}{h}\right)\left(\frac{y}{h}\right)\phi_3.\end{aligned} \tag{6-31}$$

Evaluation of the energy using this expression gives

$$\begin{aligned}W = {} & \frac{\epsilon}{6}\Big[(\phi_0 - \phi_1)^2 + (\phi_0 - \phi_1)(\phi_3 - \phi_2) + (\phi_3 - \phi_2)^2 \\ & + (\phi_0 - \phi_3)^2 + (\phi_0 - \phi_3)(\phi_1 - \phi_2) + (\phi_1 - \phi_2)^2\Big].\end{aligned} \tag{6-32}$$

The derivative of W with respect to one nodal value ϕ_0 can be evaluated as

$$\frac{\partial W}{\partial \phi_0} = \frac{\epsilon}{6}(4\phi_0 - \phi_1 - 2\phi_2 - \phi_3). \tag{6-33}$$

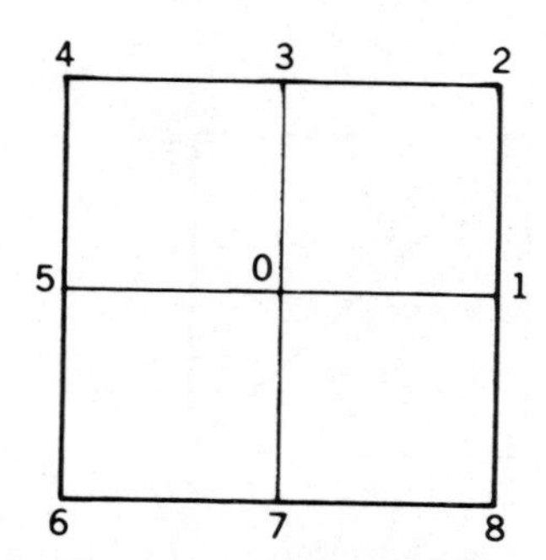

FIGURE 6-10. A typical interior node with square elements.

Using a similar expression for the other three squares shown in Fig. 6-10, the derivative of the energy for the four squares is

$$\frac{\partial W}{\partial \phi_0} = \frac{\epsilon}{6}(16\phi_0 - 2\phi_1 - 2\phi_2 - 2\phi_3 - 2\phi_4$$

$$-2\phi_5 - 2\phi_6 - 2\phi_7 - 2\phi_8). \qquad (6\text{-}34)$$

Since ϕ_0 is selected to minimize the total energy, the left side of this expression is zero and

$$\phi_0 = \frac{\phi_1 + \phi_2 + \phi_3 + \phi_4 + \phi_5 + \phi_6 + \phi_7 + \phi_8}{8}. \qquad (6\text{-}35)$$

6.5. GENERAL TRIANGULAR ELEMENTS

Next we generalize the situation from the isoceles right triangle to the general triangle, as illustrated in Fig. 6-11. An excellent account of the required analysis is given in Reference 1. We would like to use the same linear representation (6-19) that was used with the isoceles right triangle. The equations that must be satisfied for this

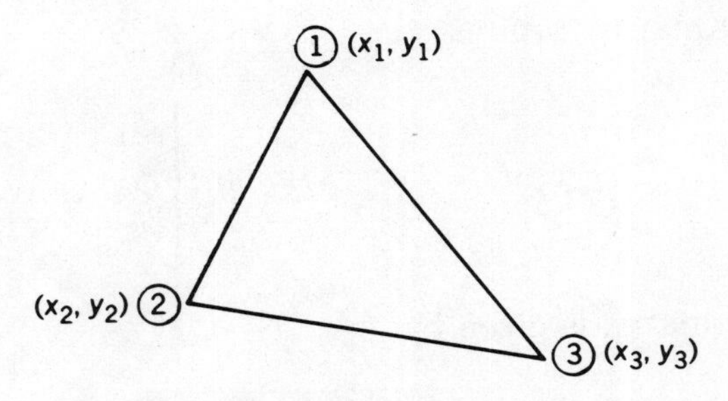

FIGURE 6-11. The general triangular element.

representation to agree with the functional values at the vertices are

$$a + bx_1 + cy_1 = \phi_1$$

$$a + bx_2 + cy_2 = \phi_2$$

$$a + bx_3 + cy_3 = \phi_3. \tag{6-36}$$

In matrix notation,

$$\begin{bmatrix} 1 & x_1 & y_1 \\ 1 & x_2 & y_2 \\ 1 & x_3 & y_3 \end{bmatrix} \begin{bmatrix} a \\ b \\ c \end{bmatrix} = \begin{bmatrix} \phi_1 \\ \phi_2 \\ \phi_3 \end{bmatrix}. \tag{6-37}$$

The determinant of the coefficients is

$$\begin{vmatrix} 1 & x_1 & y_1 \\ 1 & x_2 & y_2 \\ 1 & x_3 & y_3 \end{vmatrix} = 2A, \tag{6-38}$$

where A is the area of the triangle. Except for the special case in which the three vertices are collinear and the area is zero, the equations can be solved for a, b, and c. Then $\phi(x, y)$ can be expressed in matrix form as

$$\phi(x, y) = [\psi_1 \quad \psi_2 \quad \psi_3]\begin{bmatrix}\phi_1 \\ \phi_2 \\ \phi_3\end{bmatrix}, \tag{6-39}$$

where the ψ-matrix is given by

$$[\psi_1 \quad \psi_2 \quad \psi_3] = [1 \quad x \quad y]\begin{bmatrix}1 & x_1 & y_1 \\ 1 & x_2 & y_2 \\ 1 & x_3 & y_3\end{bmatrix}^{-1}. \tag{6-40}$$

The expression for $\phi(x, y)$ is equivalent to

$$\phi(x, y) = \sum_{k=1}^{3} \psi_k(x, y)\phi_k. \tag{6-41}$$

Evaluation of the ψ_k gives

$$\begin{aligned}
\psi_1(x, y) &= \frac{1}{2A}[(x_2y_3 - x_3y_2) + (y_2 - y_3)x + (x_3 - x_2)y] \\
\psi_2(x, y) &= \frac{1}{2A}[(x_3y_1 - x_1y_3) + (y_3 - y_1)x + (x_1 - x_3)y] \\
\psi_3(x, y) &= \frac{1}{2A}[(x_1y_2 - x_2y_1) + (y_1 - y_2)x + (x_2 - x_1)y].
\end{aligned} \tag{6-42}$$

The ψ_k are interpolation functions in the sense that

$$\psi_j(x_k, y_k) = \begin{cases} 1 & \text{if } j = k \\ 0 & \text{if } j \neq k. \end{cases} \tag{6-43}$$

To evaluate the energy integral, substitute

$$\nabla\phi(x, y) = \sum_k \phi_k \nabla\psi_k \tag{6-44}$$

into (6-22) to yield

$$W = \frac{\epsilon}{2} \sum_{j,k} \delta_j \phi_k S_{jk}, \tag{6-45}$$

where

$$S_{jk} = \iint (\nabla \psi_j) \cdot (\nabla \psi_k)\, dx\, dy. \tag{6-46}$$

The matrix coefficients S_{jk} depend only on the shape of the triangle, not on its size or location and orientation with respect to the coordinate system. For example,

$$S_{12} = \iint \left(\frac{\partial \psi_1}{\partial x} \frac{\partial \psi_2}{\partial x} + \frac{\partial \psi_1}{\partial y} \frac{\partial \psi_2}{\partial y} \right) dx\, dy, \tag{6-47}$$

which is

$$S_{12} = \frac{1}{4A} [(y_2 - y_3)(y_3 - y_1) + (x_3 - x_2)(x_1 - x_3)], \tag{6-48}$$

which can be evaluated (see Reference 1) as

$$S_{12} = -\tfrac{1}{2} \cot \theta_3, \tag{6-49}$$

where θ_3 is the interior angle at vertex 3 of the triangle. The other S_{jk} coefficients can be evaluated in a similar way, which leads to

$$S = \frac{1}{2} \begin{bmatrix} \cot \theta_2 + \cot \theta_3 & -\cot \theta_3 & -\cot \theta_2 \\ -\cot \theta_3 & \cot \theta_1 + \cot \theta_3 & -\cot \theta_1 \\ -\cot \theta_2 & -\cot \theta_1 & \cot \theta_1 + \cot \theta_2 \end{bmatrix}. \tag{6-50}$$

The energy calculated for one triangle from (6-45) is a quadratic function of the potential values at the vertices. The total energy for the entire region is the sum of the energies of each triangular element and hence is a quadratic function of all the potential values. The minimum energy is found by differentiation of this function, which results in a linear combination of potential values for each node. Thus the result is a set of linear equations that can be solved for the unknown potential values. Because the matrix of this set of equations is sparse, iterative methods of solutions normally are used.

6.6. HIGHER-ORDER INTERPOLATION WITH TRIANGLES

We wish to consider use of higher-order polynomials, but with our current approach it is clear that the algebraic computations will be complex. Clearly, we need a more systematic way to write the equations. The method we will follow involves the use of simplex coordinates, sometimes referred to as *area coordinates*. In two dimensions we normally define a point by the two spatial coordinates x and y. With simplex coordinates, we specify a point by three coordinates s_1, s_2, and s_3. These coordinates are defined locally with respect to one triangle and we use a different set to define a point in another triangle. Figure 6-12 shows a typical triangle and the s_j coordinates. Each of the three coordinates is defined as the ratio of the distance of the point from a triangle side to the distance of the opposite vertex from that side. Defined in this way it is clear that each s_j varies from 0 to 1 as the point moves from the side opposite vertex j to the vertex j. The redundancy

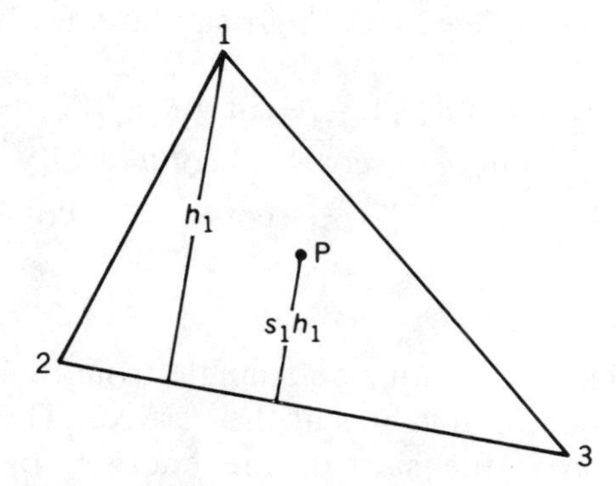

FIGURE 6-12. Simplex coordinates for the triangle.

created by using three coordinates to describe a point in two dimensions is expressed by

$$s_1 + s_2 + s_3 = 1. \tag{6-51}$$

Because each simplex coordinate is the ratio of the altitudes of two triangles with the same base, it clearly is the ratio of the areas (from which comes the other name, *area coordinate*). For example, s_1 can be expressed as

$$s_1 = \frac{\begin{vmatrix} 1 & x & y \\ 1 & x_2 & y_2 \\ 1 & x_3 & y_3 \end{vmatrix}}{\begin{vmatrix} 1 & x_1 & y_1 \\ 1 & x_2 & y_2 \\ 1 & x_3 & y_3 \end{vmatrix}}. \tag{6-52}$$

Rather than use this expression to evaluate the simplex coordinate s_1 as a function of x and y, we can see from Fig. 6-12 that if the point P is at vertex 2 or 3, s_1 is zero and if the point P is at vertex 1, s_1 is unity. Similar considerations apply to the other two simplex coordinates, s_2 and s_3. Therefore, the simplex coordinates are equal to the interpolation functions defined by (6-42). From another point of view, for linear interpolation the interpolation functions are equal to the simplex coordinates, so that

$$\begin{aligned} \psi_1 &= s_1 \\ \psi_2 &= s_2 \\ \psi_3 &= s_3. \end{aligned} \tag{6-53}$$

This suggests that higher-order interpolation formulas could be defined in terms of the simplex coordinates. Consider the quadratic interpolation expression

$$\phi(x, y) = a + bx + cy + dx^2 + exy + fy^2. \tag{6-54}$$

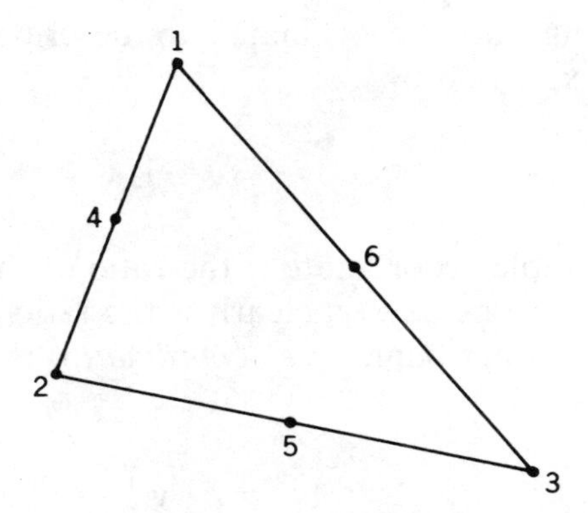

FIGURE 6-13. Nodes for the quadratic representation

Because there are six coefficients, we need six nodes, and Fig. 6-13 shows a logical way to locate these nodes. Interpolation functions for these nodes can be constructed from expressions that are quadratic in the simplex coordinates. The quadratic interpolation function ψ_1, for example, must be unity for s_1 equal to unity and zero for s_1 equal to zero and one-half. Hence

$$\psi_1 = s_1(2s_1 - 1). \tag{6-55}$$

The other five interpolation functions can be constructed in the same way and the result is

$$\begin{aligned}
\psi_1 &= s_1(2s_1 - 1)\\
\psi_2 &= s_2(2s_2 - 1)\\
\psi_3 &= s_3(2s_3 - 1)\\
\psi_4 &= 4s_1s_2\\
\psi_5 &= 4s_2s_3\\
\psi_6 &= 4s_1s_3.
\end{aligned} \tag{6-56}$$

The next higher-degree polynomial is the cubic, which has 10 terms. There are not 10 logical nodes on the triangle edges, so a node inside the triangle is used, as shown in Fig. 6-14. Similar functions can be constructed from the simplex coordinates for cubic and

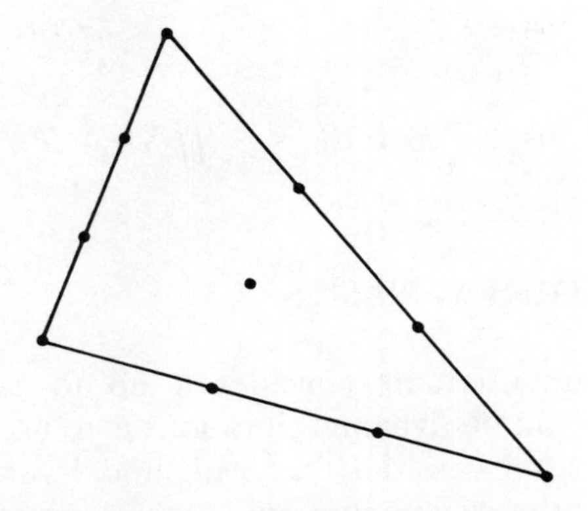

FIGURE 6-14. Nodes for the cubic representation.

higher-order interpolation in the same way as for quadratic interpolation.

6.7. NODAL EXPANSIONS AND THE WEAK FORMULATION

We have written several local approximations in the form of nodal expansions. We can write one global expansion in the form

$$\phi(x, y) = \psi_0(x, y) + \sum_{k=1}^{N} \phi_k \psi_k(x, y), \tag{6-57}$$

where $\psi_0(x, y)$ satisfies the boundary conditions, N is the total number of unknown nodal values, ϕ_k are these nodal values, and $\psi_k(x, y)$ are the interpolation formulas, which are derived from the local formulas. This global expansion can be used with the Rayleigh–Ritz approach to select the ϕ_k that minimize the total energy. This is simply another way of describing the approach that we have followed before. Use of the nodal expansion as an approximation to the weak formulation.

$$\iint \nabla\phi \cdot \nabla\psi \, dx \, dy \tag{6-58}$$

results in

$$\sum_{k=1}^{N} \phi_k \iint \nabla\psi_k \cdot \nabla\psi_j \, dx \, dy = - \iint \nabla\psi_0 \cdot \nabla\psi_j \, dx \, dy. \quad (6\text{-}59)$$

6.8. TIME-DEPENDENT VARIABLES

The finite-element situations considered up to now have been static. One way to handle dynamic situations is to use finite-element expressions for spatial variation but with nodal variables that depend on time. For the wave equation

$$\frac{\partial^2 E}{\partial t^2} = c^2 \nabla^2 E \qquad \text{with } E = 0 \text{ on the boundary.} \quad (6\text{-}60)$$

Substitution of an approximation by the nodal expansion

$$E(x, y, z, t) \approx \sum_k E_k(t) \psi_k(x, y, z) \quad (6\text{-}61)$$

gives

$$\sum_k E_k'' \psi_k \approx c^2 \sum_k E_k \nabla^2 \psi_k. \quad (6\text{-}62)$$

Multiplication by ψ_j and integration results in

$$\sum_k E_k'' \iiint \psi_j \psi_k \, dx \, dy \, dz = c^2 \sum_k E_k \iiint \psi_j \nabla^2 \psi_k, \quad (6\text{-}63)$$

which can be written through one of Green's theorems,

$$\sum_k E_k'' \iiint \psi_j \psi_k \, dx \, dy \, dz = -c^2 \sum_k E_k \iiint \nabla\psi_j \cdot \nabla\psi_k \, dx \, dy \, dz. \quad (6\text{-}64)$$

This expression is a weak formulation and has the advantage that only first spatial derivatives of the expansion functions are involved. Using a finite-difference approach with the time variable is one of several ways to solve this set of ordinary differential equations.

PROBLEMS

1. For the element that is the nonisoceles right triangle shown below, calculate the energy in terms of the nodal values.

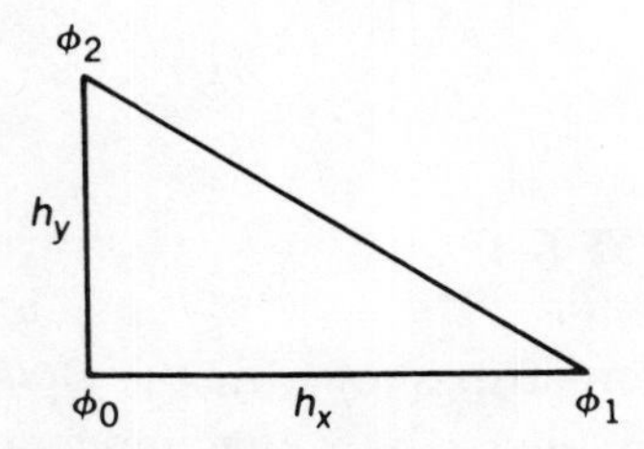

PROBLEM 6-1. Right triangular element.

2. For the square element shown with the nodal potential values in the following figure, calculate the energy for the two methods discussed in the text.

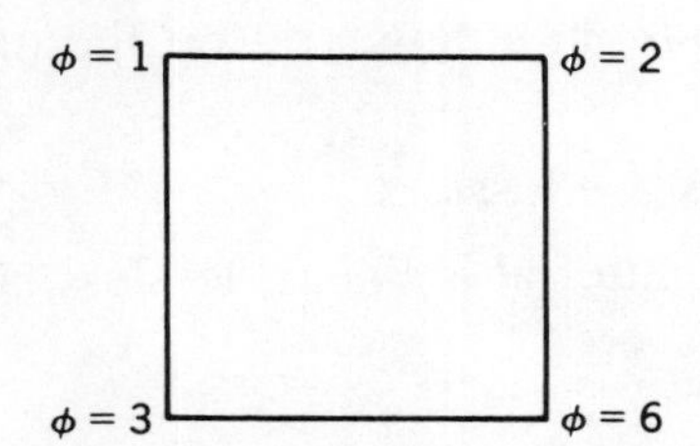

PROBLEM 6-2. Square element.

3. Calculate the energy of the nonsquare rectangular element shown below.

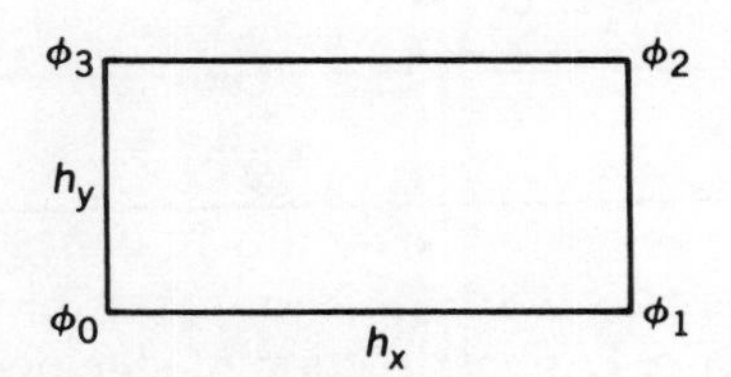

PROBLEM 6-3. Rectangular element.

4. The finite-element method is to be used with square elements and bilinear interpolation to calculate electrostatic potential. Derive the equation to be used to determine a nodal value when the node is on the boundary between two media with different permittivities.

COMPUTER PROJECT 6-1

For the shielded microstrip shown in the figure, use the finite-element method to calculate the capacitance per unit length. Use square elements with bilinear interpolation functions. Express the result in pF/m to the nearest 0.1 pF/m. Take advantage of the symmetry. The assumptions are:

1. $\epsilon_1 = 9.5\ \epsilon_0$.
2. The microstrip (conducting strip on the dielectric material) is very thin.
3. The conductors are lossless.
4. The dielectric material is linear, lossless, and isotropic.

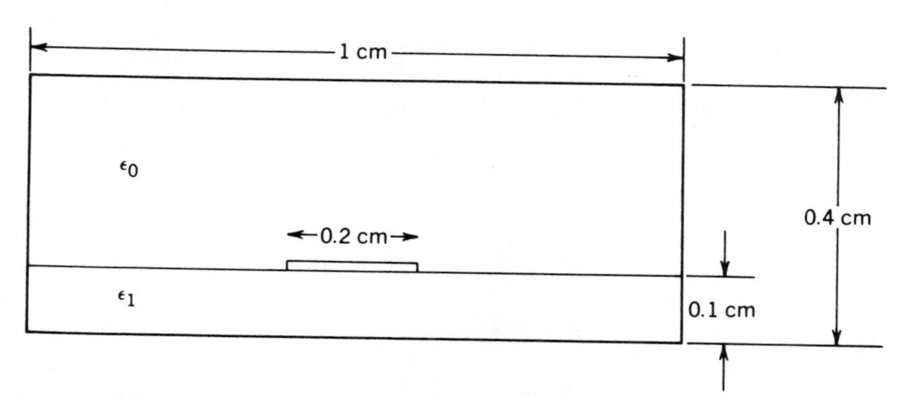

COMPUTER PROJECT 6-1. Cross section of shielded microstrip transmission line.

REFERENCES

1. P. P. Silvester and R. L. Ferrari, *Finite Elements for Electrical Engineers*, Cambridge University Press, New York, 1983.
2. C. W. Steele, *Numerical Computation of Electric and Magnetic Fields*, Van Nostrand Reinhold, New York, 1987.
3. A. J. Davies, *The Finite Element Method: A First Approach*, Oxford University Press, Oxford, 1980.
4. W. B. Bickford, *A First Course in the Finite Element Method*, Richard D. Irwin, Homewood, Ill., 1990.

CHAPTER SEVEN

Method of Moments

7.1. LINEAR OPERATORS

Electromagnetic problems usually involve solution of linear partial differential equations or integral equations. The general form of such a linear equation is the operator equation

$$\mathscr{L} f(\mathbf{R}) = g(\mathbf{R}), \tag{7-1}$$

where $\mathscr{L}$ is a linear operator, f is the function to be calculated, and g is a known function. Two electrostatic examples are

$$\nabla^2 \phi(\mathbf{R}) = -\frac{\rho(\mathbf{R})}{\epsilon_0}, \tag{7-2}$$

where the operator involves derivatives and is referred to as a *differential operator*, and

$$\iiint \frac{1}{4\pi\epsilon_0 |\mathbf{R} - \mathbf{R}'|} \rho(\mathbf{R}')\, d^3R' = \phi(\mathbf{R}), \tag{7-3}$$

where the operator is called an *integral operator*. The general integral operator can be written as

$$\iiint G(\mathbf{R}, \mathbf{R}')f(\mathbf{R}')\, d^3R' = g(\mathbf{R}), \tag{7-4}$$

where the kernel $G(\mathbf{R}, \mathbf{R}')$ is known as a *Green's function*.

7.2. APPROXIMATION BY EXPANSION IN BASIS FUNCTIONS

To solve problems such as those described by (7-1), the method of moments [1] begins by approximating the unknown function by a linear combination of known functions $\psi_k(\mathbf{R})$ in the form

$$f(\mathbf{R}) \approx \psi_0(\mathbf{R}) + \sum_{k=1}^{N} \alpha_k \psi_k(\mathbf{R}), \tag{7-5}$$

where the functions ψ_k are known as basis or expansion functions and are so selected that with appropriate values for the parameters α_k, the right side of this equation is a reasonably accurate approximation to the left side. The interpolation functions in Chapter 6 are special cases of basis functions. The function ψ_0 satisfies the same boundary conditions as the function f but otherwise can be selected arbitrarily, and the basis functions satisfy homogeneous boundary conditions. Substitution of (7-5) into (7-1) and use of the linearity of the operator results in an equation of the form

$$\sum_{k=1}^{N} \alpha_k g_k(\mathbf{R}) \approx h(\mathbf{R}), \tag{7-6}$$

where

$$g_k(\mathbf{R}) = \mathscr{L}\psi_k(\mathbf{R}) \tag{7-7}$$

and

$$h(\mathbf{R}) = g(\mathbf{R}) - \mathscr{L}\psi_0(\mathbf{R}). \tag{7-8}$$

There are two fundamental classes of basis functions, *entire-domain functions*, which span the entire domain, and *subdomain functions*, each of which is zero everywhere except in a subdomain.

7.3. DETERMINATION OF THE PARAMETERS

These two sides of (7-6) cannot in general be equal everywhere. The parameters α_k are to be determined so that the resulting linear combination is a good approximation in some sense to the desired function $f(x)$. We can impose at most N constraints from which the α_k can be calculated. One approach, called *point matching*, sets the two sides of (7-6) equal at N points $\mathbf{R}_j$, giving the set of equations

$$\sum_{k=1}^{N} \alpha_k g_k(\mathbf{R}_j) = h(\mathbf{R}_j). \tag{7-9}$$

This is a set of linear algebraic equations that can be solved to obtain the parameters α_k. Substitution of the resulting set of parameters into (7-5) completes determination of the approximate solution.

A more general approach is to multiply each side of (7-6) by one of a set of weighting functions w_j and integrate over the domain of $\mathbf{R}$ to yield

$$\sum_{k=1}^{N} \alpha_k \iiint w_j(\mathbf{R}) g_k(\mathbf{R})\, d^3R = \iiint w_j(\mathbf{R}) h(\mathbf{R})\, d^3R \tag{7-10}$$

as the set of algebraic equations to be solved for the parameters. The point-matching method can be considered to be a special case of this method by taking the weighting functions to be

$$w_j(\mathbf{R}) = \delta(\mathbf{R} - \mathbf{R}_j). \tag{7-11}$$

An important approach, known as *Galerkin's method*, uses as weighting functions the same set of functions as used for the basis functions

$$w_j(\mathbf{R}) = \psi_j(\mathbf{R}). \tag{7-12}$$

In mathematical circles, the name *method of weighted residuals* is frequently used, because if the residual Res(**R**) is defined as

$$\mathrm{Res}(\mathbf{R}) = \sum_{k=1}^{N} \alpha_k g_k(\mathbf{R}) - h(\mathbf{R}), \tag{7-13}$$

the integrated residual weighted with w_j is set equal to zero, so that

$$\iiint w_j(\mathbf{R})\mathrm{Res}(\mathbf{R})\, d^3R = 0, \tag{7-14}$$

which clearly is equivalent to (7-10). The set of equations (7-10) can be written in matrix form as

$$A\alpha = B, \tag{7-15}$$

where

$$A_{jk} = \iiint w_j(\mathbf{R}) g_k(\mathbf{R})\, d^3R \tag{7-16}$$

and

$$B_j = \iiint w_j(\mathbf{R}) h(\mathbf{R})\, d^3R. \tag{7-17}$$

7.4. DIFFERENTIAL OPERATORS

Solution of problems involving differential operators corresponds to the solution of differential equations. As a simple one-dimensional example, consider

$$\frac{d^2}{dx^2} f(x) = -x^2, \tag{7-18}$$

with the boundary conditions

$$f(0) = 0 \qquad \text{and} \qquad f(1) = 0. \tag{7-19}$$

These boundary conditions results in ψ_0 being zero. Basis functions with continuous derivatives at least to the first order are desirable to avoid delta functions. A suitable family of basis functions for this problem is

$$\psi_k(x) = x - x^{k+1} \qquad \text{for } k = 1, 2, \ldots, N \tag{7-20}$$

and using the Galerkin approach, we take the weighting functions as the same functions

$$w_j(x) = x - x^{j+1}. \tag{7-21}$$

Then, for example, $N = 2$ gives the approximation

$$f(x) \approx \alpha_1(x - x^2) + \alpha_2(x - x^3). \tag{7-22}$$

Substitution into (7-9) yields for the matrix coefficients

$$\begin{aligned}
A_{11} &= \int_0^1 (x - x^2)(2)\,dx = \tfrac{1}{3} \\
A_{12} &= \int_0^1 (x - x^2)(6x)\,dx = \tfrac{1}{2} \\
A_{21} &= \int_0^1 (x - x^3)(2)\,dx = \tfrac{1}{2} \\
A_{22} &= \int_0^1 (x - x^3)(6x)\,dx = \tfrac{4}{5} \\
B_1 &= \int_0^1 (x - x^2)(x^2)\,dx = \tfrac{1}{20} \\
B_2 &= \int_0^1 (x - x^3)(x^2)\,dx = \tfrac{1}{12}.
\end{aligned} \tag{7-23}$$

The matrix equation for the parameters thus is

$$\begin{bmatrix} \frac{1}{3} & \frac{1}{2} \\ \frac{1}{2} & \frac{4}{5} \end{bmatrix} \begin{bmatrix} \alpha_1 \\ \alpha_2 \end{bmatrix} = \begin{bmatrix} \frac{1}{20} \\ \frac{1}{12} \end{bmatrix}, \tag{7-24}$$

whose solution yields

$$\begin{bmatrix} \alpha_1 \\ \alpha_2 \end{bmatrix} = \begin{bmatrix} -\frac{1}{10} \\ \frac{1}{6} \end{bmatrix}, \tag{7-25}$$

and thus the approximation is

$$f(x) \approx -\tfrac{1}{10}(x - x^2) + \tfrac{1}{6}(x - x^3). \tag{7-26}$$

7.5. INTEGRAL OPERATORS

The second class of operators, the *integral operators*, involve weighted integrals of the unknown function in the form

$$g(\mathbf{R}) = \iiint G(\mathbf{R}, \mathbf{R}')f(\mathbf{R}')\, d^3R', \tag{7-27}$$

For this operator (7-8) becomes

$$g_k(\mathbf{R}) = \iiint G(\mathbf{R}, \mathbf{R}')\psi_k(\mathbf{R}')\, d^3R'. \tag{7-28}$$

The resulting matrix coefficients are

$$A_{jk} = \iiiiiint w_j(\mathbf{R})G(\mathbf{R}, \mathbf{R}')\psi_k(\mathbf{R}')\, d^3R'\, d^3R \tag{7-29}$$

and

$$B_j = \iiint w_j(\mathbf{R})h(\mathbf{R})\, dR. \tag{7-30}$$

With the Galerkin method these matrix coefficients are

$$A_{jk} = \iiiiiint \psi_j(\mathbf{R})G(\mathbf{R}, \mathbf{R}')\psi_k(\mathbf{R}')\, d^3R'\, d^3R \tag{7-31}$$

and

$$B_j = \iiint \psi_j(\mathbf{R})h(\mathbf{R})\, dR. \tag{7-32}$$

7.6. PULSE FUNCTIONS

One of the most widely used type of subdomain function is the unit pulse function, which is equal to unity over one subdomain and zero elsewhere. If the typical subdomain is denoted by S_k,

$$\psi_k(\mathbf{R}) = \begin{cases} 1 & \text{if } \mathbf{R} \text{ is in } S_k \\ 0 & \text{elsewhere} \end{cases} \tag{7-33}$$

Pulse functions are especially useful with integral operators, where they frequently are used for both basis and weighting functions. With pulse basis functions, (7-28) becomes

$$g_k(\mathbf{R}) = \iiint_{S_k} G(\mathbf{R}, \mathbf{R}')\, d^3R', \tag{7-34}$$

and with pulse functions for both basis and weighting functions, the matrix coefficients given by (7-31) and (7-32) become

$$A_{jk} = \iiint_{S_j} \iiint_{S_k} G(\mathbf{R}, \mathbf{R}')\, d^3R'\, d^3R \tag{7-35}$$

and

$$B_j = \iiint_{S_j} h(\mathbf{R})\, d^3R. \tag{7-36}$$

Because with this approach we have divided the source region into a set of "finite elements" with simple approximations to the unknown function over each element, this approach sometimes is considered to be a part of the finite-element method. The difference between the present approach and the finite element method discussed earlier is that here we usually are dividing just the source region and not the entire domain.

7.7. PARALLEL-PLATE CAPACITOR IN TWO DIMENSIONS

We now consider a two-dimensional capacitor that is the cross section of a parallel-plate transmission line as shown by Fig. 7-1.

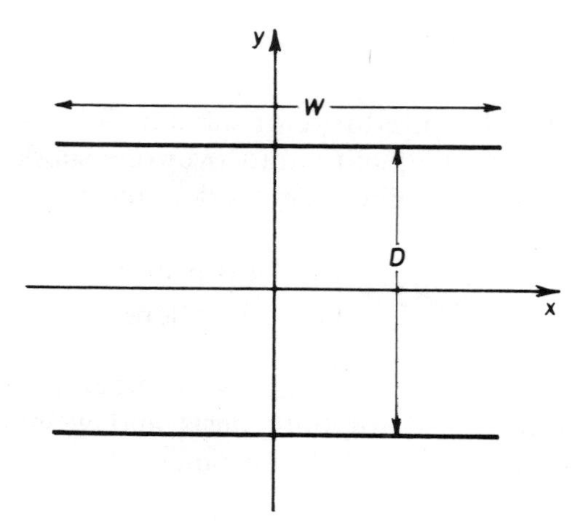

FIGURE 7-1. Cross section of parallel plate transmission line.

Denote the width by W and the separation by D. In two dimensions, the relation equivalent to (7-3) is

$$\phi(x, y) = -\frac{1}{2\pi\epsilon_0}\int\int \rho(x', y')\ln\sqrt{(x - x')^2 + (y - y')^2}\, dx'\, xy'. \tag{7-37}$$

Take the potential on the plate at $y' = D/2$ as $+1$ and the potential on the plate at $y' = -D/2$ as -1. We wish to determine the capacitance per unit length of this capacitor by first solving the integral equation (7-37) to determine the charge densities on the plates and then integrating the densities to determine the total charges, from which the capacitance can be determined. If the plates are assumed to be very thin, the integration is over the one-dimensional region of the plates and can be written as two line integrals, one for $y' = D/2$ and one for $y' = -D/2$,

$$\begin{aligned}\phi(x, y) = &-\frac{1}{2\pi\epsilon_0}\int_{-0.5W}^{0.5W} \rho_s(x', D/2)\ln\sqrt{(x - x')^2 + (y - D/2)^2}\, dx' \\ &-\frac{1}{2\pi\epsilon_0}\int_{-0.5W}^{0.5W} \rho_s(x', -D/2)\ln\sqrt{(x - x')^2 + (y + D/2)^2}\, dx',\end{aligned} \tag{7-38}$$

where now the charge density ρ_s corresponds to a surface charge density. Because

$$\rho_s(x', -D/2) = -\rho_s(x', D/2), \tag{7-39}$$

(7-38) can be written as

$$\phi(x, y) = \frac{1}{2\pi\epsilon_0}\int_{-0.5W}^{0.5W} \rho_s(x', D/2)\ln\sqrt{\frac{(x-x')^2 + (y+D/2)^2}{(x-x')^2 + (y-D/2)^2}}\, dx'. \tag{7-40}$$

Because of the symmetry with respect to x,

$$\rho_s(-x', D/2) = \rho_s(x', D/2) \tag{7-41}$$

the equation can be written as

$$\phi(x, y) = \int_0^{0.5W} \rho_s(x', D/2)G(x, y, x')\, dx'. \tag{7-42}$$

where

$$G(x, y, x') = \frac{1}{2\pi\epsilon_0}\left[\ln\sqrt{\frac{(x'-x)^2 + (y+D/2)^2}{(x'-x)^2 + (y-D/2)^2}} + \ln\sqrt{\frac{(x'+x)^2 + (y+D/2)^2}{(x'+x)^2 + (y-D/2)^2}}\right]. \tag{7-43}$$

Because the potential on the upper plate is unity, we can set $y = D/2$ and get

$$1 = \int_0^{0.5W} \rho_s(x')G(x, x')\, dx' \qquad \text{for } 0 \le x \le 0.5W \tag{7-44}$$

and G can be written

$$G(x, x') = \frac{1}{2\pi\epsilon_0}\left[\ln\sqrt{\frac{(x'-x)^2 + D^2}{(x'-x)^2}} + \ln\sqrt{\frac{(x'+x)^2 + D^2}{(x'+x)^2}}\right], \tag{7-45}$$

where we have simplified the notation for ρ_s and G. Equations (7-44) and (7-45) describe the specific equation to be solved.

One approach to the solution of the integral equation is to use pulse functions for the basis functions and point matching at the midpoints of the subintervals to derive the equations for the parameters. First, divide the interval $(0, W/2)$ into N subintervals of length

$$h = \frac{W}{2N}, \tag{7-46}$$

and then the kth subinterval is

$$S_k = (kh - h, kh). \tag{7-47}$$

An approximation in pulse functions leads to

$$\rho_s(x) \approx \sum_{k=1}^{N} \alpha_k P_k(x), \tag{7-48}$$

where

$$P_k(x) = \begin{cases} 1 & \text{if } kh - h \le x \le kh \\ 0 & \text{elsewhere.} \end{cases} \tag{7-49}$$

If (7-48) is substituted into (7-42) and if the two sides are point matched at the midpoint,

$$x_j = jh - 0.5h, \tag{7-50}$$

the equation becomes

$$\sum_{k=1}^{N} \alpha_k \int_{kh-h}^{kh} G(x_j, x')\, dx' = 1 \qquad \text{for } j = 1, 2, \ldots, N, \tag{7-51}$$

where from (7-45),

$$G(x_j, x') = \frac{1}{2\pi\epsilon_0}\left[\ln\sqrt{\frac{(x'-x_j)^2 + D^2}{(x'-x_j)^2}} + \ln\sqrt{\frac{(x'+x_j)^2 + D^2}{(x'+x_j)^2}}\right]. \tag{7-52}$$

The relations (7-51) and (7-52) define the set of algebraic equations that must be solved to determine the α_k parameters. Because the potential difference between the two plates is 2, the capacitance, which is

$$C = \frac{Q}{\Delta\phi}, \tag{7-53}$$

becomes

$$C = \tfrac{1}{2}\int_{-W/2}^{W/2} \rho_s(x')\, dx' \tag{7-54}$$

and because of the symmetry

$$C = \int_{0}^{W/2} \rho_s(x')\, dx'. \tag{7-55}$$

Because ρ_s is approximated by pulse functions, the integral in (7-55) can be approximated by

$$C = \sum_{k=1}^{N} \alpha_k h. \tag{7-56}$$

It is convenient to rewrite (7-51) as

$$\sum_{k=1}^{N} \alpha_k h \frac{1}{h} \int_{kh-h}^{kh} G(x_j, x')\, dx' = 1 \tag{7-57}$$

and then take $\alpha_k h$ as the unknown parameters. Because α_k is the charge density in the kth subdomain, $\alpha_k h$ is approximately the charge in that subdomain. Then if (7-57) is written in matrix form as

$$A\alpha h = 1, \tag{7-58}$$

the matrix coefficients are

$$A_{jk} = \frac{1}{h} \int_{kh-h}^{kh} G(x_j, x')\, dx' \tag{7-59}$$

and

$$B_j = 1. \tag{7-60}$$

Substitution of (7-52) into (7-59) and use of (7-50) leads to

$$A_{jk} = \frac{1}{2\pi\epsilon_0} \left[\frac{1}{h} \int_{kh-h}^{kh} \ln \sqrt{\frac{(x' - jh + 0.5h)^2 + D^2}{(x' - jh + 0.5h)^2}}\, dx' + \frac{1}{h} \int_{kh-h}^{kh} \ln \sqrt{\frac{(x' + jh - 0.5h)^2 + D^2}{(x' + jh - 0.5h)^2}}\, dx' \right]. \tag{7-61}$$

With the variable change

$$\beta = \frac{x' - jh + 0.5h}{D} \tag{7-62}$$

in the first integral, and in the second integral

$$\beta = \frac{x' + jh - 0.5h}{D}, \tag{7-63}$$

(7-61) becomes

$$A_{jk} = \frac{1}{2\pi\epsilon_0\gamma}\left[\int_{\gamma(k-j-0.5)}^{\gamma(k-j+0.5)} \ln\sqrt{\frac{\beta^2 + 1^2}{\beta^2}}\, d\beta + \int_{\gamma(k+j-1.5)}^{\gamma(k+j-0.5)} \ln\sqrt{\frac{\beta^2 + 1^2}{\beta^2}}\, d\beta\right], \tag{7-64}$$

where

$$\gamma = \frac{h}{D} = \frac{W/D}{2N}. \tag{7-65}$$

Equation (7-64) can be expressed as

$$A_{jk} = \frac{1}{2\pi\epsilon_0\gamma}[F_1(\gamma k - \gamma j + 0.5\gamma) - F_1(\gamma k - \gamma j - 0.5\gamma) + F_1(\gamma k + \gamma j - 0.5\gamma) - F_1(\gamma k + \gamma j - 1.5\gamma)], \tag{7-66}$$

where F_1 is an indefinite integral

$$F_1(\beta) = \int \ln\sqrt{\frac{\beta^2 + 1^2}{\beta^2}}\, d\beta, \tag{7-67}$$

which can be evaluated as

$$F_1(\beta) = \beta \ln\sqrt{\frac{\beta^2 + 1^2}{\beta^2}} + \tan^{-1}\beta. \tag{7-68}$$

For an alternative solution, the Galerkin method uses for the testing function w_j the same function as the basis function. With pulse functions for both basis and testing function, instead of (7-51) the basic equation becomes

$$\sum_{k=1}^{N} \alpha_k \int_{jh-h}^{jh} \int_{kh-h}^{kh} G(x, x')\, dx'\, dx = \int_{jh-h}^{jh} 1\, dx \qquad \text{for } j = 1, 2, \ldots, N, \tag{7-69}$$

where G is given by (7-45). The right side of the equation is equal to h and thus the equation can be written as

$$\sum_{k=1}^{N} \alpha_k h \frac{1}{h^2} \int_{jh-h}^{jh} \int_{kh-h}^{kh} G(x, x')\, dx'\, dx = 1 \qquad \text{for } j = 1, 2, \ldots, N. \tag{7-70}$$

If as before the unknowns are taken as $\alpha_k h$, the matrix coefficients are

$$A_{jk} = \frac{1}{h^2} \int_{jh-h}^{jh} \int_{kh-h}^{kh} G(x, x')\, dx'\, dx \tag{7-71}$$

and

$$B_j = 1. \tag{7-72}$$

By making changes of variable similar to (7-62) and (7-63), the right side of (7-71) can be evaluated as

$$\begin{aligned} A_{jk} = \frac{1}{2\pi\epsilon_0 \gamma^2} [& F_2(\gamma k - \gamma j + \gamma) - 2F_2(\gamma k - \gamma j) \\ & + F_2(\gamma k - \gamma j - \gamma) + F_2(\gamma k + \gamma j) \\ & - 2F_2(\gamma k + \gamma j - \gamma) + F_2(\gamma k + \gamma j - 2\gamma)], \end{aligned} \tag{7-73}$$

where F_2, which is an indefinite integral of F_1, is given by

$$F_2(\beta) = \begin{cases} 0 & \text{if } \beta = 0 \\ \dfrac{\beta^2 - 1}{2} \ln \sqrt{\beta^2 + 1} - \dfrac{\beta^2}{2} \ln \sqrt{\beta^2} + \beta \tan^{-1} \beta & \text{if } \beta \neq 0. \end{cases} \tag{7-74}$$

The following table shows the results of using these two methods to calculate the capacitance for the particular case $W = D$.

	Capacitance (pF/m)	
N	Point Matching	Pulse Weighting
1	17.1098	17.7084
2	17.8677	18.1861
4	18.2858	18.4444
8	18.5059	18.5851
16	18.6188	18.6583

The correct value can be shown by further computation and extrapolation to be approximately 18.73 pF/m. Figure 7-2 shows the charge distribution for pulse weighting with $N = 16$.

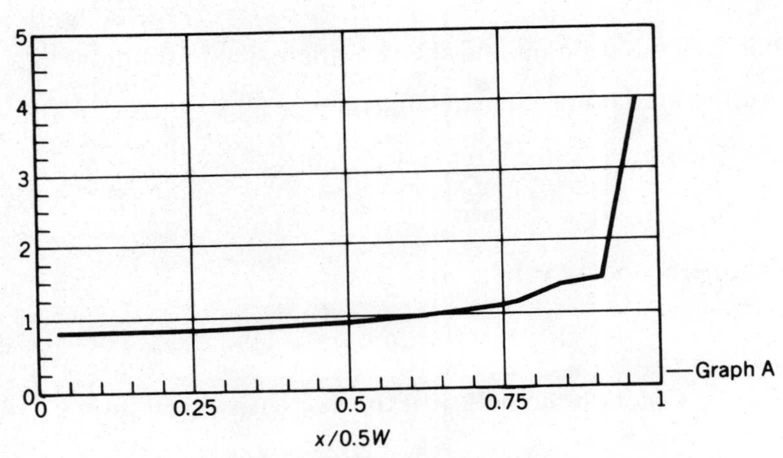

FIGURE 7-2. Charge distribution.

PROBLEMS

1. Use the method of moments to solve

$$\frac{d^2}{dx^2}f(x) = -6x,$$

with the boundary conditions

$$f(0) = 0 \qquad \text{and} \qquad f(1) = 0.$$

Use $N = 1$ and take the one basis function as

$$f_1(x) = x - x^2.$$

Evaluate α_1 for the two weighting functions

$$w_1 = x - x^2$$

and

$$w_1 = 1.$$

By integration, the true solution is

$$f(x) = x - x^3.$$

Plot the true solution and the two approximate solutions.

2. Use the method of moments to solve

$$\frac{d^2}{dx^2}f(x) = -x^2,$$

with the boundary conditions

$$f(0) = 0 \qquad \text{and} \qquad f(1) = 0.$$

Use the Galerkin method with the two basis functions

$$f_1(x) = x - x^2$$

and

$$f_2(x) = x - x^3.$$

Compare the values for $x = 0.5$ of the true solution and the approximate solution.

COMPUTER PROJECT 7-1

Calculate the capacitance per unit length of the unshielded parallel-plate transmission line using the Galerkin method with pulse functions and the parameter ratio

$$\frac{W}{D} = 2.$$

Carry out a second solution using the Galerkin method with just one basis function

$$\psi(x) = \frac{1}{\sqrt{1 - (x/0.5W)^2}}.$$

For the second solution, evaluate the required integrals numerically. Compare the two solution methods.

COMPUTER PROJECT 7-2

Calculate the capacitance with respect to infinity of a thin flat square plate with 1-m sides. Divide the square into M^2 squares and use the moment method with two-dimensional pulse functions for basis functions. Do point matching at the center of each small square. Evaluate the integrals involving the Green's function numerically.

REFERENCE

1. R. F. Harrington, *Field Computation by Moment Methods* (Reprint Edition), R. E. Krieger, Malabar, Fla., 1987; Original Edition, 1968.

CHAPTER EIGHT

Scattering Solutions by Method of Moments

8.1. FUNDAMENTAL SCATTERING EQUATIONS

The solution of scattering problems is one of the major applications of the method of moments. In such a problem, a wave called the *incident wave* strikes an object, which we will usually assume is a perfectly conducting body. This wave causes a current to flow on the object's surface, and in turn, this current causes a wave called the *scattered wave*, which can be expressed as

$$\overline{\mathbf{E}}^s(\mathbf{R}) = \iint G(\mathbf{R}, \mathbf{R}')\overline{\mathbf{J}}_s(\mathbf{R}')\, d^2R', \tag{8-1}$$

where the integration is over the surface of the body and J_s is the surface current density (in A/m). If the body is a perfect conductor, the tangential value of the total E field is zero on the surface so that

$$\overline{\mathbf{E}}_t^i + \overline{\mathbf{E}}_t^s = 0. \tag{8-2}$$

Combination of these two equations yields

$$\left(\iint G(\mathbf{R}, \mathbf{R}')\overline{\mathbf{J}}_s(\mathbf{R}')\, d^2R'\right)_t = -\overline{\mathbf{E}}_t^i(R). \tag{8-3}$$

This is the fundamental equation that we wish to solve with the method of moments.

First, the Green's function must be derived. The Green's function derived here is for scattering in free space. The waveguide iris discussed in Chapter 9 offers another example of scattering, with the Green's function changed by the waveguide boundary conditions. The scattered electric field can be expressed in terms of the magnetic vector potential as

$$\overline{\mathbf{E}}^s = \frac{1}{j\omega\mu_0\epsilon_0}\left[\nabla(\nabla \cdot \overline{\mathbf{A}}) + \beta^2\overline{\mathbf{A}}\right], \tag{8-4}$$

where in the usual notation $\beta^2 = \omega^2\mu\epsilon$. In turn, the magnetic vector potential is given by

$$\overline{\mathbf{A}}(\mathbf{R}) = \iint \frac{\mu_0\overline{\mathbf{J}}_s(\mathbf{R}')}{4\pi}\frac{e^{-j\beta D}}{D}\,d^2R', \tag{8-5}$$

where the distance D is

$$D = |\mathbf{R} - \mathbf{R}'| = \sqrt{(x - x')^2 + (y - y')^2 + (z - z')^2} \tag{8-6}$$

in rectangular coordinates. Combination of (8-4) and (8-5) yields

$$\overline{\mathbf{E}}^s = \frac{1}{j4\pi\omega\epsilon_0}\iint\left[\nabla\left(\nabla \cdot \overline{\mathbf{J}}_s(\mathbf{R}')\frac{e^{-j\beta D}}{D}\right) + \beta^2\overline{\mathbf{J}}_s(\mathbf{R}')\frac{e^{-j\beta D}}{D}\right]d^2R'. \tag{8-7}$$

The various forms of this basic equation can be solved in the usual way with the moment method [1]. In contrast to most of the relatively simple examples we have considered in earlier chapters, however, the integrals that are be evaluated to define the matrix coefficients must be evaluated numerically. The result is that for scattering and other difficult problems, the majority of the computer run time is devoted to "filling the matrix," that is, calculation of the matrix terms.

8.2. SCATTERING OF A PLANE WAVE OFF A THIN CYLINDER

For a specific example, consider the uniform plane wave traveling in the x-direction:

$$E_z^i(x) = E_m e^{-j\beta x}. \tag{8-8}$$

This wave encounters a thin cylinder (wire) of radius a and length L, whose axis lies on the z-axis. If the wire is sufficiently thin, then approximately on the surface of the cylinder,

$$E_z^i(x) \approx E_m. \tag{8-9}$$

For this situation, the current flows only in the z-direction and is uniform around the cylinder, and hence has the same effect as a current concentrated on the z-axis. Thus the z-component of the vector potential can be written in terms of the total current as

$$A_z(r, z) = \frac{\mu_0}{4\pi} \int_0^L \frac{e^{-j\beta D}}{D} I(z')\, dz', \tag{8-10}$$

where in the cylindrical coordinates used in this equation

$$D = \sqrt{r^2 + (z - z')^2}. \tag{8-11}$$

On the surface of the cylinder

$$A_z(a, z) = \frac{\mu_0}{4\pi} \int_0^L \frac{e^{-j\beta D}}{D} I(z')\, dz', \tag{8-12}$$

where now

$$D = \sqrt{a^2 + (z - z')^2}. \tag{8-13}$$

Because D depends only upon $z - z'$, it is convenient to write

$$K(z - z') = \frac{e^{-j\beta D}}{D} \tag{8-14}$$

and (8-12) can be written as

$$A_z(a, z) = \frac{\mu_0}{4\pi} \int_0^L K(z - z')I(z')\, dz'. \quad (8\text{-}15)$$

Thus, on the surface of the cylinder, (8-2) and (8-4) reduce to

$$-j4\pi\omega\epsilon_0 E_m = \left(\frac{\partial^2}{\partial z} + \beta^2\right) \int_0^L K(z - z')I(z')\, dz'. \quad (8\text{-}16)$$

Several approaches have been used with this equation. The direct one is to apply the operator to the integrand and get

$$-j4\pi\omega\epsilon_0 E_m = \int_0^L G(z, z')I(z')\, dz', \quad (8\text{-}17)$$

where the Green's function is

$$G(z,z') = \left(\frac{\partial^2}{\partial z} + \beta^2\right) K(z - z'), \quad (8\text{-}18)$$

which is one form of Pocklington's equation. Performing the differentiation leads to the explicit result

$$G(z, z') = e^{-j\beta D}\left[\frac{2j\beta}{D^2} + \frac{2 + a^2\beta^2}{D^3} - \frac{3j\beta a^2}{D^4} - \frac{3a^2}{D^5}\right]. \quad (8\text{-}19)$$

The current can be seen to be symmetrical around the midpoint of the wire, and thus if the origin is redefined so that the equation (8-17) becomes

$$-j4\pi\omega\epsilon_0 E_m = \int_{-L/2}^{L/2} G(z, z')I(z')\, dz', \quad (8\text{-}20)$$

the fact that I is an even function results in

$$-j4\pi\omega\epsilon_0 E_m = \int_0^{L/2} G(z, z')I(z')\, dz', \quad (8\text{-}21)$$

where now the Green's function is

$$G(z, z') = \left(\frac{\partial^2}{\partial z} + \beta^2\right)[K(z - z') + K(z + z')]. \quad (8\text{-}22)$$

With this Green's function, the integral equation (8-17) can be solved with the moment method. The integration required to obtain the matrix coefficients must be performed numerically. Useful results can be obtained using pulse functions for the basis functions, but as the number N of basis functions is increased, these integrals do not converge well and the behavior of the solution as a function of N is not too smooth. Basis functions that correspond to a smoother approximation for the current produce better results.

Another approach can be taken to improve the convergence of the solution process. Equation (8-16) is a second-order differential equation, and thus the integral can be written as

$$\int_0^{L/2} G(z, z')I(z')\, dz' = \frac{1}{\beta^2} j4\pi\omega\epsilon_0 E_m + B\sin(\beta z) + C\cos(\beta z), \quad (8\text{-}23)$$

which is one form of Hallén's equation. This integral equation can be solved with better convergence properties than Pocklington's equation. The coefficient B is seen to be zero because of the symmetry, and C must be determined when the current is calculated. If $N - 1$ basis functions are used and N conditions are satisfied, then solution of the system of N equations yields the $N - 1$ parameters that define the approximation to the current. One can make use of the fact that the current is zero at the end of the wire.

Butler and Wilton compare in [2] various numerical techniques that have been used with this scattering problem.

8.3. SCATTERING OF A PLANE WAVE OFF A THICK CYLINDER

If the cylinder is thick, which means that the radius a is not small compared to the wavelength of the incident wave, several of the approximations made in the preceding section are not valid. First,

(8-9) is not a good approximation. If we write the incident field in cylindrical coordinates (r, ϕ, z), then

$$x = r\cos\phi$$
$$y = r\sin\phi, \tag{8-24}$$

and the incident field can be written from (8-8) as

$$E_z^i(r,\phi) = E_m e^{-j\beta r\cos\phi}. \tag{8-25}$$

Second, the current depends upon the angle ϕ around the axis and (8-10) becomes

$$A_z(r,\phi,z) = \frac{a\mu_0}{4\pi}\int_0^L\int_0^{2\pi} e^{-j\beta D}\frac{1}{D}J_z(a,\phi',z')\,d\phi'\,dz', \tag{8-26}$$

where

$$D = \sqrt{r^2 - 2ar\cos(\phi-\phi') + a^2 + (z-z')^2}. \tag{8-27}$$

The electric field corresponding to this vector potential is, as before,

$$E_z^s(r,\phi,z) = \frac{1}{j\omega\epsilon_0\mu_0}\left(\frac{\partial^2}{\partial z^2} + \beta^2\right)A_z(r,\phi,z). \tag{8-28}$$

Combination of these equations and setting $r = a$ gives the desired integral equation,

$$-E_m e^{-j\beta a\cos\phi} = \frac{a}{j4\pi\omega\epsilon_0}\int_0^L\int_0^{2\pi}\left(\frac{\partial^2}{\partial z^2} + \beta^2\right)e^{-j\beta D} \times\frac{1}{D}J_s(\phi',z')\,d\phi'\,dz', \tag{8-29}$$

which corresponds to (8-19) and where now

$$D = \sqrt{4a^2 \sin^2 \frac{\phi - \phi'}{2} + (z - z')^2}\,. \tag{8-30}$$

In addition, the current on the ends of the cylinder, which was previously neglected, now must be included.

PROBLEMS

1. Use integration by parts on equation (8-17) to derive another form of Pocklington's equation

$$-j4\pi\omega\epsilon_0 E_m = \int_0^L K(z - z')\left[I''(z') + \beta^2 I(z')\right] dz' + AK(z) + BK(z - L)$$

where A and B are constants determined by the current I. What are A and B?

2. Using the fact that $\beta = 2\pi/\lambda$ and the change of variables $\alpha = \beta z$ to show that the nature of the solution to Pocklington's equation depends upon only the two parameters a/λ and L/λ.

3. Derive the integral equation corresponding to (8-20) if the incident wave is not normal to the wire.

COMPUTER PROJECT 8-1

Write a computer program to solve with the moment method the integral equation described by equations (8-21) and (8-22). Use pulse functions for basis functions and point match at the center of the pulses. Run the program for the following parameters:

$$\frac{a}{\lambda} = 0.01 \qquad \frac{L}{\lambda} = 0.25$$

and plot the solution for the current at the center of the wire as a function of $1/N$, where N is the number of basis functions. Do the solutions converge well as N is increased?

REFERENCES

1. R. C. Hansen (ed.), *Moment Methods in Antennas and Scattering*, Artech House, Norwood, Mass., 1990.
2. Chalmers M. Butler and D. R. Wilton, "Analysis of various numerical techniques applied to thin-wire scatterers," *IEEE Transactions on Antennas and Propagation*, AP-23, 1975.

CHAPTER NINE

Spectral Analysis with Fourier Series and Fourier Integral

9.1. BASIC FOURIER SERIES RELATIONS

In practice the most difficult task in the application of the method of moments usually is determination of the Green's function. Only in a few situations, such as the parallel-plate capacitor in unbounded space, can analytic expressions for the Green's function be derived. In principle, as shown in Chapter 7, determination of the Green's function is straightforward. If the function to be found is the solution of an operator equation, as in

$$\mathscr{L}[f(\mathbf{R})] = g(\mathbf{R}), \tag{9-1}$$

with the boundary condition

$$f(\mathbf{R}) = 0 \text{ on the boundary}, \tag{9-2}$$

the Green's function $G(\mathbf{R}, \mathbf{R}')$ is the solution of

$$\mathscr{L}[G(\mathbf{R}, \mathbf{R}')] = \delta(\mathbf{R} - \mathbf{R}') \qquad \text{with the same boundary condition.} \tag{9-3}$$

Sometimes, the defining equation (9-3) can be solved directly in closed analytic form. For many situations, however, solutions for the Green's function as an infinite series or as an integral transform are required. Waveguide and shielded microstrip examples are good illustrations of such a situation and will be considered later in the chapter.

9.2. EXAMPLE INVOLVING LAPLACE'S EQUATION

First, consider a simple example, with a variable $E(x, y)$ that satisfies Laplace's equation inside the rectangular region shown in Fig. 9-1. The function is zero on three sides and equal to $f(x)$ on the upper side. If for each value of y the function is expanded in a Fourier series,

$$E(x, y) = \sum_{n=1}^{N} A_n(y) \sin \frac{n\pi x}{a}. \tag{9-4}$$

If each term in this series satisfies Laplace's equation,

$$\frac{\partial^2 A_n}{\partial y^2} - \left(\frac{n\pi}{a}\right)^2 A_n = 0. \tag{9-5}$$

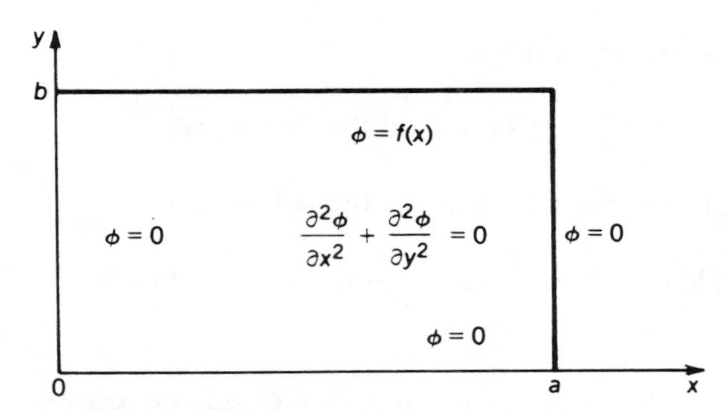

FIGURE 9-1. A two-dimensional potential region.

To satisfy the boundary conditions, the appropriate solution is

$$A_n(y) = a_n \sinh \frac{n\pi y}{a}, \tag{9-6}$$

and the series is

$$E(x, y) = \sum_{n=1}^{N} a_n \sinh \frac{n\pi y}{a} \sin \frac{n\pi x}{a}. \tag{9-7}$$

This series satisfies the zero boundary condition on three of the sides. To satisfy the boundary condition on the upper side, the coefficients a_n are determined in the usual Fourier series manner as

$$a_n \sinh \frac{n\pi b}{a} = \frac{2}{a} \int_0^a \sin \frac{n\pi x'}{a} f(x')\, dx' \tag{9-8}$$

and the series is

$$E(x, y) = \sum_{n=1}^{N} \frac{\sinh(n\pi y/a)}{\sinh(n\pi b/a)} \sin \frac{n\pi x}{a} \frac{2}{a} \int_0^a \sin \frac{n\pi x'}{a} f(x')\, dx'. \tag{9-9}$$

This expression can be written as

$$E(x, y) = \int_0^a G(x, y, x') f(x')\, dx', \tag{9-10}$$

where the Green's function is

$$G(x, y, x') = \frac{2}{a} \sum_{n=1}^{N} \frac{\sinh(n\pi y/a)}{\sinh(n\pi b/a)} \sin \frac{n\pi x}{a} \sin \frac{n\pi x'}{a}. \tag{9-11}$$

If the boundary conditions are nonzero on the other three sides, three other similar series must be added to the solution.

9.3. INDUCTIVE IRIS IN A WAVEGUIDE

An example of a scattering problem that is also a good illustration of the use of Fourier series techniques is the inductive iris in a waveguide. With the usual notation for a rectangular waveguide, the TE_{n0} mode has an E-field defined by

$$E_{yn}(x, z) = \sin\frac{n\pi x}{a} e^{\pm\gamma_n z}, \tag{9-12}$$

where the propagation constant γ_n satisfies

$$\gamma_n^2 = \left(\frac{n\pi}{a}\right)^2 - k^2, \qquad \text{where } k^2 = \omega^2\mu\epsilon. \tag{9-13}$$

Usual waveguide practice results in

$$\frac{\pi}{a} < k < \frac{2\pi}{a}, \tag{9-14}$$

and thus γ_1 is imaginary and γ_n is real for all values of n greater than 1. Now assume that a dominant mode wave defined by

$$E^i(x, z) = \sin\frac{\pi x}{a} e^{-\gamma_1 z} \tag{9-15}$$

is traveling in the positive z-direction and at $z = 0$ there is a thin diaphragm (iris) with the cross section shown in Fig. 9-2. If this iris is a conductor, current will flow in the y-direction and create a

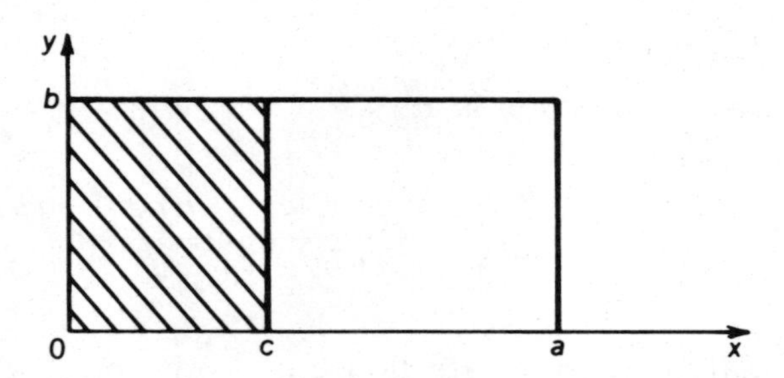

FIGURE 9-2. Inductive iris in a rectangular waveguide.

scattered wave. If the iris is a perfect conductor, the total E-field on the iris will be zero. To determine the scattered field, we need to find the current on the iris. First we need an equation that relates the scattered field to the current. Although the incident E-field is a sinusoidal function of x, the resulting current is not. The current density can be expanded in a Fourier series of the form

$$J_y(x) = \sum_{n=1}^{\infty} c_n \sin\frac{n\pi x}{a}, \tag{9-16}$$

where the coefficients c_n are given by

$$c_n = \frac{2}{a}\int_0^a \sin\frac{n\pi x'}{a} J_y(x')\,dx'. \tag{9-17}$$

For z slightly less than 0, the magnetic field is given by

$$H_x^s(x, 0-) = -\tfrac{1}{2}J_y(x). \tag{9-18}$$

Because for each mode the E and H fields are related by

$$E_{yn}^s = \frac{j\omega\mu}{\gamma_n} H_{xn}^s, \tag{9-19}$$

the scattered E-field is derived from (9-16) using (9-18) and (9-19) as

$$E_y^s(x, 0) = \sum_{n=1}^{\infty}\left(-\frac{j\omega\mu c_n}{2\gamma_n}\right)\sin\frac{n\pi x}{a} \tag{9-20}$$

for $z = 0$, and for $z < 0$

$$E_y^s(x, z) = \sum_{n=1}^{\infty}\left(-\frac{j\omega\mu c_n}{2\gamma_n}\right)\sin\frac{n\pi x}{a} e^{\gamma_n z}. \tag{9-21}$$

Substitution of (9-17) and rearrangement yields

$$E_y^s(x, z) = -\int_0^c \sum_{n=1}^{\infty} \frac{j\omega\mu}{a\gamma_n} \sin\frac{n\pi x}{a} e^{\gamma_n z} \sin\frac{n\pi x'}{a} J_y(x')\, dx'. \quad (9\text{-}22)$$

In order that the total field on the iris be zero,

$$E_y^s(x, 0) = -\sin\frac{\pi x}{a}, \quad (9\text{-}23)$$

which leads to the integral equation

$$\int_0^c G(x, x') J_y(x')\, dx' = \sin\frac{\pi x}{a} \qquad \text{for } 0 \le x \le c, \quad (9\text{-}24)$$

where the Green's function is

$$G(x, x') = \sum_{n=1}^{\infty} \frac{j\omega\mu}{a\gamma_n} \sin\frac{n\pi x}{a} \sin\frac{n\pi x'}{a}. \quad (9\text{-}25)$$

The moment-method solution based on the approximation

$$J_y(x) = \sum_{j=1}^{N} \alpha_j \psi_j(x) \quad (9\text{-}26)$$

and the weighting functions $w_i(x)$ leads to

$$\sum_{j=1}^{N} \alpha_j \int_0^c \int_0^c w_i(x) G(x, x') \psi_j(x')\, dx'\, dx = \int_0^c w_i(x) \sin\frac{\pi x}{a}\, dx, \quad (9\text{-}27)$$

which can be written

$$\sum_{j=1}^{N} \alpha_j \sum_{n=1}^{\infty} \frac{j\omega\mu}{a\gamma_n} \left[\int_0^c w_i(x) \sin\frac{n\pi x}{a}\, dx\right] \left[\int_0^c \sin\frac{n\pi x'}{a} \psi_j(x')\, dx'\right]$$

$$= \int_0^c w_i(x) \sin\frac{\pi x}{a}\, dx. \quad (9\text{-}28)$$

For the Galerkin method with pulse functions, use N subintervals of length

$$h = \frac{c}{N}, \tag{9-29}$$

and the first integral is

$$\int_0^c w_i(x)\sin\frac{n\pi x}{a}\,dx = \int_{ih-h}^{ih} \sin\frac{n\pi x}{a}\,dx, \tag{9-30}$$

which easily can be evaluated as

$$\int_0^c w_i(x)\sin\frac{n\pi x}{a}\,dx = \frac{a}{n\pi}\left[\cos\left(\frac{n\pi h}{a}(i-1)\right) - \cos\left(\frac{n\pi h}{a}(i)\right)\right]. \tag{9-31}$$

This integral depends on n and i and the parameter h/a, which can be expressed as

$$\frac{h}{a} = \frac{c/a}{N}. \tag{9-32}$$

The other two integrals are of the same form and thus it is convenient to write (9-28) as

$$\sum_{j=1}^{N} \alpha_j \sum_{n=1}^{\infty} \frac{j\omega\mu}{a\gamma_n}\left(\frac{a}{n\pi}\right)^2 F(n,i)F(n,j) = \frac{a}{n\pi}F(1,i). \tag{9-33}$$

where

$$F(n,i) = \cos\left(\frac{n\pi h}{a}(i-1)\right) - \cos\left(\frac{n\pi h}{a}(i)\right). \tag{9-34}$$

It is convenient to rearrange (9-33) as

$$\sum_{j=1}^{N} \alpha_j \frac{\omega\mu a}{\pi} \sum_{n=1}^{\infty} \frac{j}{n^2 a\gamma_n} F(n,i)F(n,j) = F(1,i), \tag{9-33}$$

and then by introducing the normalized set of parameters

$$X_j = \alpha_j \frac{\omega m a}{\pi} \tag{9-34}$$

the equation becomes

$$\sum_{j=1}^{N} X_j \sum_{n=1}^{\infty} \frac{j}{n^2 a\gamma_n} F(n,i)F(n,j) = F(1,i). \tag{9-33}$$

The coefficient $a\gamma_n$ is found from (9-13) as

$$a\gamma_n = \sqrt{(n\pi)^2 - (ka)^2}\,. \tag{9-34}$$

Because for large n this coefficient is proportional to only the first power of n, the series for the Green's function can be seen from (9-25) not to converge well. The two integrals, however, each introduce a factor of $1/n$, and therefore the terms in the series in (9-33) go to zero as $1/n^3$ and therefore the series converges well.

The dominant term in the scattered (reflected) wave is the $n = 1$ term in (9-22) and thus the reflection coefficient at $z = 0$ is

$$\rho = -\frac{j\omega\mu}{a\gamma_1} \int_0^c \sin\frac{\pi x}{a} J_y(x)\,dx. \tag{9-35}$$

Using the basis function expansion we used,

$$J_y(x) = \sum_{j=1}^{N} \alpha_j P_j(x), \tag{9-36}$$

the reflection coefficient can be expressed as

$$\rho = -\frac{j\omega\mu}{a\gamma_1} \int_0^c \sin\frac{\pi x}{a} \sum_{j=1}^{N} \alpha_j P_j(x)\,dx, \tag{9-37}$$

which is

$$\rho = -\sum_{j=1}^{N} \alpha_j \frac{j\omega\mu}{a\gamma_1} \int_0^c \sin\frac{\pi x}{a} P_j(x)\, dx. \tag{9-38}$$

In terms of our previous notation,

$$\rho = -\frac{j}{a\gamma_1} \sum_{j=1}^{N} X_j F(1, j). \tag{9-39}$$

In circuit terms, the effect of the iris on the dominant mode can be represented as a shunt impedance across a transmission line. The normalized shunt impedance can be shown to be

$$\frac{Z_s}{Z_0} = -\frac{1+\rho}{2\rho}. \tag{9-40}$$

The right side of this equation can be shown to be an imaginary number, leading to the iris being referred to as an *inductive iris*.

9.4. BASIC FOURIER INTEGRAL RELATIONS

In Section 9.3 we demonstrated the usefulness of Fourier series representations for variables defined on finite domains, as in the case of waveguides and shielded microstrip circuits. For variables defined on infinite domains, the series are not useful and we must turn to the Fourier transform to use similar methods. The Fourier transform $\tilde{F}(\alpha)$ of a function $F(x)$ is defined by the integral

$$\tilde{F}(\alpha) = \int_{-\infty}^{+\infty} e^{-j\alpha x} F(x)\, dx. \tag{9-41}$$

The function $F(x)$ can be recovered from its transform $\tilde{F}(\alpha)$ by the integral

$$F(x) = \frac{1}{2\pi} \int_{-\infty}^{+\infty} e^{+j\alpha x} \tilde{F}(\alpha)\, d\alpha. \tag{9-42}$$

These relations are analogous to the Fourier series. The function $F(x)$ now is defined over an infinite domain rather than the finite domain in the case of the series. The function $\tilde{F}$ of the continuous variable α replaces the discrete set of Fourier coefficients in the Fourier series case.

9.5. FOURIER INTEGRAL SOLUTION

Consider now the situation where the side walls in the structure shown in Fig. 9-1 are removed to leave the region shown in Fig. 9-3. Suppose that the function $E(x, y)$ is zero for $y = 0$ and is equal to $f(x)$ for $y = b$. The transform of both sides of Laplace's equation is

$$\left(-\alpha^2 + \frac{\partial^2}{\partial y^2}\right)\tilde{E}(\alpha, y) = 0, \tag{9-43}$$

which has solutions that can be expressed as linear combinations of hyperbolic sines and cosines. Because of the first boundary condition, the sinh function will be used and thus

$$\tilde{E}(\alpha, y) = c(\alpha)\sinh \alpha y. \tag{9-44}$$

Use of the second boundary condition to evaluate $c(\alpha)$ leads to

$$\tilde{E}(\alpha, y) = \frac{\sinh \alpha y}{\sinh \alpha b}\tilde{f}(\alpha). \tag{9-45}$$

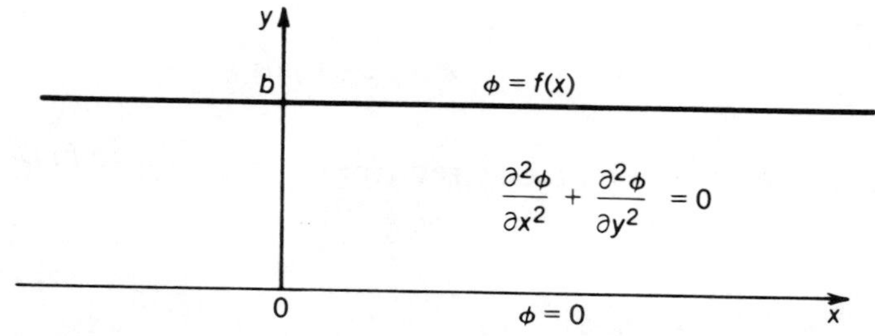

FIGURE 9-3. A two-dimensional potential region without sides.

Taking the inverse transform gives

$$E(x, y) = \int_{-\infty}^{+\infty} G(x, y, x') f(x')\, dx', \tag{9-46}$$

where the Green's function is

$$G(x, y, x') = \frac{1}{2\pi} \int_{-\infty}^{+\infty} e^{j\alpha(x-x')} \frac{\sinh \alpha y}{\sinh \alpha b}\, d\alpha. \tag{9-47}$$

The integral in this expression probably must be evaluated numerically.

PROBLEM

1. For the inductive iris, derive expressions for A_{ij} and B_i for the basis and weighting functions

$$\psi_j(x) = \sin \frac{j\pi x}{a}$$

$$w_i(x) = \sin \frac{i\pi x}{a}.$$

COMPUTER PROJECT 9-1

Write a program to implement the Galerkin with pulse functions solution of the inductive iris problem. Use the value for *ka* of 4.5, and solve for various values of c/a and graph ρ as a function of c/a from 0 to 1.

REFERENCES

1. D. Mirshekar-Syahkal, *Spectral Domain Method for Microwave Integrated Circuits*, John Wiley & Sons, New York, 1990.
2. Craig Scott, *The Spectral Domain Method in Electromagnetics*, Artech House, Norwood, Mass., 1989.

CHAPTER TEN

Spectral Analysis of Microstrip Transmission Lines

10.1. MICROSTRIP TRANSMISSION LINE

Two versions of microstrip transmission line will be analyzed here, shielded and unshielded, shown in Figs. 10-1 and 10-2, respectively. Because of the two values of permittivity, microstrip transmission lines cannot support a TEM mode of transmission. At sufficiently low frequencies, however, the dominant mode is approximately TEM, and quasistatic or quasi-TEM calculations based on the steady-state values of inductance L and capacitance C per unit length give useful results. Most analyses do not calculate L directly. Instead, the capacitance C_0 that results from assuming that $\epsilon_1 = \epsilon_0$ is determined, along with the capacitance C_1 that corresponds to the true value of ϵ_1. These two values of capacitance are then used to determine the transmission line parameters such as phase velocity. As the frequency is increased, the quasi-TEM results become less accurate, and eventually the "full-wave" solution that results from using Maxwell's equations must be determined.

10.2. QUASI-TEM SPECTRAL ANALYSIS OF SHIELDED MICROSTRIP TRANSMISSION LINE

Consider now the quasi-TEM solution for the shielded microstrip transmission line. The capacitance per unit length is to be calcu-

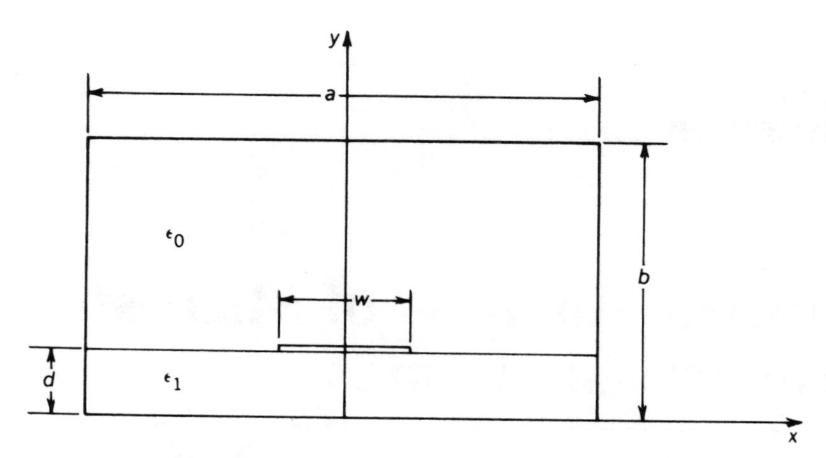

FIGURE 10-1. Shielded microstrip transmission line.

lated. In regions where ϵ is constant, the potential satisfies Laplace's equation

$$\frac{\partial^2 \phi}{\partial x^2} + \frac{\partial^2 \phi}{\partial y^2} = 0. \tag{10-1}$$

The potential clearly is an even function of x and on the side walls satisfies the boundary conditions

$$\phi(-0.5a, y) = 0 \tag{10-2}$$

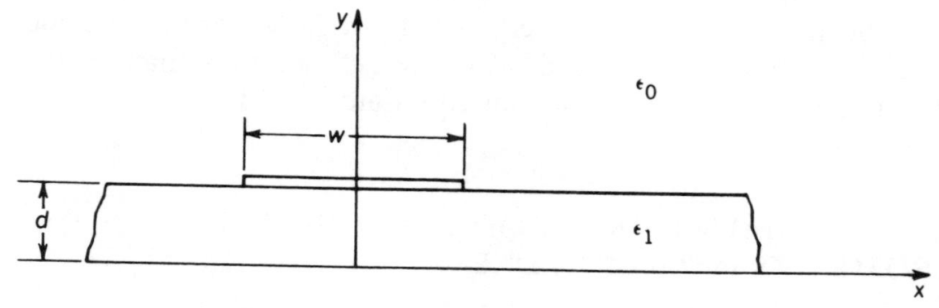

FIGURE 10-2. Unshielded microstrip transmission line.

and

$$\phi(0.5a, y) = 0. \tag{10-3}$$

The potential in the lower region ($y < h$), which will be denoted by ϕ_1, also satisfies the boundary condition

$$\phi_1(x, 0) = 0. \tag{10-4}$$

The separation of variables approach leads to the solution

$$\phi_1(x, y) = \sum{}' c_n \sinh \frac{n\pi y}{a} \cos \frac{n\pi x}{a}, \tag{10-5}$$

where the prime indicates summation over odd positive values of n $(1, 3, 5, \ldots)$. The potential in the upper region ($h < y$), which will be denoted by ϕ_0, also satisfies the boundary condition

$$\phi_1(x, b) = 0, \tag{10-6}$$

and the separation of variables approach leads to the solution

$$\phi_0(x, y) = \sum{}' d_n \sinh \frac{n\pi(y - b)}{a} \cos \frac{n\pi x}{a}. \tag{10-7}$$

The charge density, which is at the interface, can be expanded in a Fourier series

$$\rho_s(x) = \sum{}' f_n \cos \frac{n\pi x}{a}, \tag{10-8}$$

where the coefficients can be written as

$$f_n = \frac{2}{a} \int_{-0.5a}^{+0.5a} \rho_s(x) \cos \frac{n\pi x}{a}\, dx. \tag{10-9}$$

The coefficients c_n and d_n will be determined by satisfying term by term the conditions at the interface between the regions ($y = h$). The continuity of the potential

$$\phi_0(x, h) = \phi_1(x, h) \tag{10-10}$$

leads to

$$c_n \sinh \frac{n\pi h}{a} = d_n \sinh \frac{n\pi(h-b)}{a} \tag{10-11}$$

and the interface condition

$$-\epsilon_0 \frac{\partial}{\partial y}\phi_0 + \epsilon_1 \frac{\partial}{\partial y}\phi_1 = \rho_s \tag{10-12}$$

at $y = h$ leads to

$$-\epsilon_0 d_n \frac{n\pi}{a} \cosh \frac{n\pi}{a}(h-b) + \epsilon_1 c_n \frac{n\pi}{a} \cosh \frac{n\pi}{a} h = f_n. \tag{10-13}$$

Combination of these equations leads to

$$\phi(x,h) = \sum{}' g_n \cos \frac{n\pi x}{a} \int_{-0.5a}^{+0.5a} \cos \frac{n\pi x'}{a} \rho_s(x')\, dx', \tag{10-14}$$

where

$$g_n = \frac{2/n\pi}{-\epsilon_0 \coth(n\pi/a)(h-b) + \epsilon_1 \coth(n\pi/a)h}, \tag{10-15}$$

which can be written as

$$\phi(x,h) = \int_{-0.5a}^{+0.5a} G(x,x')\rho_s(x')\, dx', \tag{10-16}$$

where the Green's function is

$$G(x,x') = \sum{}' g_n \cos \frac{n\pi x}{a} \cos \frac{n\pi x'}{a}. \tag{10-17}$$

If the upper shield is removed, or alternatively, taking $b = \infty$, the term g_n becomes

$$g_n = \frac{2/n\pi}{\epsilon_0 + \epsilon_1 \coth(n\pi/a)h}. \tag{10-18}$$

If the potential on the strip is taken to be unity, (10-16) gives the integral equation that determines the charge density as

$$\int_{-0.5a}^{+0.5a} G(x, x')\rho_s(x')\,dx' = 1. \tag{10-19}$$

10.3. QUASI-TEM SPECTRAL ANALYSIS OF UNSHIELDED MICROSTRIP TRANSMISSION LINE

In terms of the Fourier transform

$$\tilde{\phi}(\alpha, y) = \int_{-\infty}^{+\infty} e^{-j\alpha x}\phi(x, y)\,dx, \tag{10-20}$$

Laplace's equation becomes

$$\left(\frac{\partial^2}{\partial y^2} - \alpha^2\right)\tilde{\phi}(\alpha, y) = 0. \tag{10-21}$$

The solutions to this equation can be expressed in terms of exponential functions or in terms of hyperbolic functions. To satisfy the boundary condition that the potential is zero at $y = 0$, the solution for $y < h$ is

$$\tilde{\phi}_1(\alpha, y) = c(\alpha)\sinh \alpha y, \tag{10-22}$$

and for $h < y$ the solution is

$$\tilde{\phi}_0(\alpha, y) = d(\alpha)e^{-\alpha y}. \tag{10-23}$$

Continuity of the potential at $y = h$ gives

$$c(\alpha) = \frac{d(\alpha)e^{-\alpha h}}{\sinh \alpha h}, \tag{10-24}$$

and the interface condition

$$-\epsilon_0 \frac{\partial}{\partial y}\phi_0 + \epsilon_1 \frac{\partial}{\partial y}\phi_1 = \rho_s \tag{10-25}$$

at $y = h$ leads to

$$-\epsilon_0 d(\alpha)\alpha + \epsilon_1 c(\alpha)\alpha \cosh \alpha h = \rho_s. \tag{10-26}$$

Combination of these equations leads to

$$\phi(x, h) = \int_{-\infty}^{+\infty} G(x, x')\rho_s(x')\, dx', \tag{10-27}$$

where

$$G(x, x') = \frac{1}{2\pi}\int_{-\infty}^{+\infty} e^{j(x-x')}\frac{1}{\alpha(\epsilon_0 + \epsilon_1 \coth \alpha h)}\, d\alpha. \tag{10-28}$$

Use of the symmetry with respect to α gives

$$G(x, x') = \frac{1}{\pi}\int_0^{\infty} \cos \alpha(x - x')\frac{1}{\alpha(\epsilon_0 + \epsilon_1 \coth \alpha h)}\, d\alpha. \tag{10-29}$$

10.4. COMPARISON OF FOURIER SERIES AND FOURIER INTEGRAL SOLUTIONS

Physically, one would expect that if the shield over the shielded microstrip were far from the strip that the solution would be approximately the same as the solution of the unshielded microstrip. Mathematically, this means that the Fourier series and the

Fourier integral solutions would be related. One way to see this is to examine the Green's function for the shielded microstrip.

$$G(x, x') = \frac{1}{\pi}\int_0^\infty \cos \alpha(x - x') \frac{1}{\alpha(\epsilon_0 + \epsilon_1 \coth \alpha h)}\, d\alpha. \quad (10\text{-}30)$$

The integral can be evaluated numerically by dividing the region from 0 to ∞ into subintervals of length 2δ and using the midpoint integration formula. The points of the intervals are $n\delta$, where n is an odd positive integer. This evaluation of the integral results in

$$G(x, x') = \frac{1}{\pi}\sum{}' \cos n\delta(x - x') \frac{1}{n\delta(\epsilon_0 + \epsilon_1 \coth n\,\delta h)} 2\delta. \quad (10\text{-}31)$$

Comparison with the Fourier series solution

$$G(x, x') = \sum{}' g_n \cos \frac{n\pi x}{a} \cos \frac{n\pi x'}{a}, \quad (10\text{-}32)$$

where

$$g_n = \frac{2/n\pi}{\epsilon_0 + \epsilon_1 \coth(n\pi/a)h}, \quad (10\text{-}33)$$

shows that the solutions almost agree if the subinterval in the numerical evaluation of the integral satisfies

$$\delta = \frac{\pi}{a}. \quad (10\text{-}34)$$

If the width a is large, the interval 2δ is small enough that the numerical evaluation of the integral should be accurate. One difference remains because the trigonometric term in the integral is $\cos \alpha(x - x')$, whereas the corresponding term is $\cos \alpha x \cos \alpha x'$. Clearly, in the limit the term involving $\sin \alpha x \sin \alpha x'$ must sum to zero.

PROBLEM

1. Derive the result in Section 10.2 for no top cover ($b = \infty$).

REFERENCES

1. D. Mirshekar-Syahkal, *Spectral Domain Method for Microwave Integrated Circuits*, John Wiley & Sons, New York, 1990.
2. Craig Scott, *The Spectral Domain Method in Electromagnetics*, Artech House, Norwood, Mass., 1989.

CHAPTER ELEVEN

Spectral Analysis of Microstrip Circuits

11.1. MICROSTRIP CIRCUITS

In this chapter we consider the spectral analysis of microstrip circuits. The basic structure is similar to that of the microstrip transmission line considered in Chapter 6. The substrate sits on a ground plane, but the metallization on the top of the substrate no longer is restricted to a constant-width line but can have an arbitrary shape in the $x - z$ plane. The methods used are similar to those of Chapter 10, with the difference that the lateral variation is in two dimensions rather than one. As in Chapter 10, we consider both shielded and unshielded microstrip. We first perform quasistatic analysis for both types of circuit and then do full-wave analysis for the unshielded microstrip. The full-wave analysis of the shielded microstrip circuit is similar except that Fourier series rather than Fourier integrals are used. The quasi-static analysis calculates the capacitance with respect to the ground plane of an arbitrarily shaped metal figure on the substrate, whereas the full-wave analysis determines the relation between current in the microstrip and the electric field across the substrate.

Problems and computer projects are not included at the end of this chapter, because writing a computer program to implement these methods is a major effort and the results are too close to the level of current research in this field.

11.2. QUASI-TEM SPECTRAL ANALYSIS OF SHIELDED MICROSTRIP CIRCUITS

Consider now the quasi-TEM solution for the shielded microstrip circuit, which is a three-dimensional problem. In regions where ϵ is constant, the potential satisfies Laplace's equation

$$\frac{\partial^2 \phi}{\partial x^2} + \frac{\partial^2 \phi}{\partial y^2} + \frac{\partial^2 \phi}{\partial z^2} = 0. \tag{11-1}$$

The separation of variables approach leads to the solution in the lower region of

$$\phi_1(x, y, z) = \sum_{m,n} c_{mn} \sinh(u_{mn} y) \sin \frac{n\pi x}{a} \sin \frac{n\pi z}{c}, \tag{11-2}$$

where

$$u_{mn} = \sqrt{\left(\frac{m\pi}{a}\right)^2 + \left(\frac{n\pi}{c}\right)^2}. \tag{11-3}$$

The potential in the upper region ($h < y$), which will be denoted by ϕ_0, can be written

$$\phi_0(x, y, z) = \sum_{m,n} d_{mn} \sinh(u_{mn}(y - b)) \sin \frac{n\pi x}{a} \sin \frac{n\pi z}{c}. \tag{11-4}$$

The charge density, which is at the interface, can be expanded in a Fourier series

$$\rho_s(x, z) = \sum_{m,n} f_{mn} \sin \frac{m\pi x}{a} \sin \frac{n\pi z}{c}, \tag{11-5}$$

where the coefficients can be written as

$$f_{mn} = \frac{4}{ac} \int_0^c \int_0^a \rho_s(x, z) \sin \frac{m\pi x}{a} \sin \frac{n\pi z}{c}\, dx\, dz. \tag{11-6}$$

The coefficients c_{mn} and d_{mn} will be determined by satisfying term by term the conditions at the interface between the regions ($y = h$). The continuity of the potential

$$\phi_0(x, h, z) = \phi_1(x, h, z) \tag{11-7}$$

leads to

$$c_{mn} \sinh(u_{mn} y) = d_n \sinh(u_{mn}(y - b)), \tag{11-8}$$

and the interface condition

$$-\epsilon_0 \frac{\partial}{\partial y} \phi_0 + \epsilon_1 \frac{\partial}{\partial y} \phi_1 = \rho_s \tag{11-9}$$

leads to

$$-\epsilon_0 d_{mn} u_{mn} \cosh u_{mn}(h - b) + \epsilon_1 c_{mn} u_{mn} \cosh u_{mn} h = f_{mn}. \tag{11-10}$$

Combination of these equations leads to

$$\phi(x, h, z) = \int_0^c \int_0^a G(x, z, x', z') \rho_s(x', z')\, dx'\, dz', \tag{11-11}$$

where the Green's function is

$$G(x, z, x', z') = \sum_{m,h} g_{mn} \sin \frac{m\pi x}{a} \sin \frac{n\pi z}{c} \sin \frac{n\pi x'}{a} \sin \frac{n\pi z'}{c} \tag{11-12}$$

and where

$$g_n = \frac{(1/u_{mn})(4/ac)}{-\epsilon_0 \coth u_{mn}(h - b) + \epsilon_1 \coth u_{mn} h}. \tag{11-13}$$

The solution using the moment method then proceeds in the usual manner.

11.3. QUASI-TEM SPECTRAL ANALYSIS OF UNSHIELDED MICROSTRIP CIRCUITS

Now consider the unshielded microstrip circuit. As before, at low frequencies the electric potential is governed by Laplace's equation,

$$\frac{\partial^2 \phi}{\partial x^2} + \frac{\partial^2 \phi}{\partial y^2} + \frac{\partial^2 \phi}{\partial z^2} = 0. \qquad (11\text{-}14)$$

In terms of the Fourier transform of ϕ, defined by

$$\tilde{\phi}(\alpha, y, \beta) = \iint e^{-j\alpha x} e^{-j\beta z} \phi(x, y, z)\, dx\, dz, \qquad (11\text{-}15)$$

where unless otherwise indicated limits of integration are $-\infty$ and ∞, Laplace's equation becomes

$$\left(-\alpha^2 + \frac{\partial^2}{\partial y^2} - \beta^2\right)\tilde{\phi} = 0, \qquad (11\text{-}16)$$

which is

$$\left(\frac{\partial^2}{\partial y^2} - u^2\right)\tilde{\phi} = 0, \qquad (11\text{-}17)$$

where

$$u = \sqrt{\alpha^2 + \beta^2}. \qquad (11\text{-}18)$$

In the dielectric material, where $y < h$, the appropriate solution is

$$\tilde{\phi}_1(\alpha, y, \beta) = c(\alpha, \beta) \sinh uy \qquad (11\text{-}19)$$

and above the interface

$$\tilde{\phi}_0(\alpha, y, \beta) = d(\alpha, \beta) e^{-uy}. \qquad (11\text{-}20)$$

Because the potential is continuous across the interface at $y = h$ and therefore its transform also is

$$c(\alpha, \beta)\sinh uh = d(\alpha, \beta)e^{-uh}. \tag{11-21}$$

Because the interface condition on the normal D leads to

$$-\epsilon_0 \frac{\partial}{\partial y}\phi_0 + \epsilon_1 \frac{\partial}{\partial y}\phi_1 = \rho_s, \tag{11-22}$$

the transform satisfies

$$-\epsilon_0 \frac{\partial}{\partial y}\tilde{\phi}_0 + \epsilon_1 \frac{\partial}{\partial y}\tilde{\phi}_1 = \tilde{\rho}_s. \tag{11-23}$$

Substitution of the two expressions for $\tilde{\phi}$ gives

$$\epsilon_0 u d e^{-uh} + \epsilon_1 u c \cosh uh = \tilde{\rho}_s. \tag{11-24}$$

Combination of these equations gives for the transform potential at the interface

$$\tilde{\phi}(\alpha, h, \beta) = \frac{1}{\epsilon_0 + \epsilon_1 \coth uh}\frac{1}{u}\tilde{\rho}_s. \tag{11-25}$$

The potential ϕ can then be found as the inverse transform

$$\phi(x, h, z) = \left(\frac{1}{2\pi}\right)^2 \iint e^{j\alpha x} e^{j\beta z} \frac{1}{u(\epsilon_0 + \epsilon_1 \coth uh)}\tilde{\rho}_s(\alpha, h, \beta)\, d\alpha\, d\beta. \tag{11-26}$$

Substitution of

$$\tilde{\rho}_s(\alpha, \beta) = \iint e^{-j\alpha x'} e^{-j\beta z'} \rho_s(x', z') dx'\, dz' \tag{11-27}$$

and interchange of the order of integration yields

$$\phi(x, h, z) = \iint G(x, z, x', z')\rho_s(x', z')\, dx'\, dz', \quad (11\text{-}28)$$

where the Green's function is

$$G(x, z, x', z') = \left(\frac{1}{2\pi}\right)^2 \iint e^{j\alpha(x-x')} e^{j\beta(z-z')} \times \frac{1}{u(\epsilon_0 + \epsilon_1 \coth uh)}\, d\alpha\, d\beta. \quad (11\text{-}29)$$

With the variable changes

$$\begin{aligned} \alpha &= u \cos \psi \\ \beta &= u \sin \psi \\ x - x' &= r \cos \theta \\ z - z' &= r \sin \theta \end{aligned} \quad (11\text{-}30)$$

the Green's function becomes

$$G(x, z, x', z') = \left(\frac{1}{2\pi}\right)^2 \int_0^\infty \int_0^{2\pi} e^{jru \cos(\theta - \psi)} \times \frac{1}{u(\epsilon_0 + \epsilon_1 \coth uh)} u\, d\theta\, d\gamma. \quad (11\text{-}31)$$

Because the Bessel function of zero order can be represented by the integral

$$J_0(x) = \frac{1}{2\pi} \int_0^{2\pi} e^{-jx \cos \theta}\, d\theta, \quad (11\text{-}32)$$

the integral with respect to θ is

$$\int_0^{2\pi} e^{-jru \cos(\theta - \psi)}\, d\theta = 2\pi J_0(ru), \quad (11\text{-}33)$$

and the expression for the Green's function becomes

$$G(r) = \frac{1}{2\pi}\int_0^\infty \frac{J_0(ru)}{\epsilon_0 + \epsilon_1 \coth uh}\,du. \tag{11-34}$$

Because the variable r is given by

$$r = \sqrt{(x - x')^2 + (z - z')^2}, \tag{11-35}$$

the Green's function depends only on the distance between the primed and unprimed points. The integral equation can be solved with the moment method in the usual way, but numerical integration clearly will be required to determine the Green's function. A useful approach is to evaluate and store values for this function at a discrete set of values of λ and when the function is needed for other values to interpolate between the discrete values.

11.4. FULL-WAVE SOLUTION OF UNSHIELDED MICROWAVE CIRCUITS

We shall derive the integral equation by first using the fact that the electric field satisfies the wave equation. In a region where μ and ϵ are constant, the E-field satisfies

$$\left(\frac{\partial^2}{\partial x^2} + \frac{\partial^2}{\partial y^2} + \frac{\partial^2}{\partial z^2} + k^2\right)\mathbf{E} = 0, \tag{11-36}$$

where as usual

$$k^2 = \omega^2\mu\epsilon. \tag{11-37}$$

The two-dimensional Fourier transform of $\mathbf{E}$, defined by

$$\tilde{\mathbf{E}}(\alpha, y, \beta) = \iint e^{-j(\alpha x + \beta z)}\mathbf{E}(x, y, z)\,dx\,dz, \tag{11-38}$$

satisfies

$$\left(-\alpha^2 + \frac{\partial^2}{\partial y^2} - \beta^2 + k^2\right)\mathbf{E} = 0. \tag{11-39}$$

This equation can be written as

$$\left(\frac{\partial^2}{\partial y^2} - u^2\right)\mathbf{E} = 0, \tag{11-40}$$

where

$$u = \sqrt{\alpha^2 + \beta^2 - k^2}. \tag{11-41}$$

This is a second-order ordinary differential equation with respect to y and has solutions that can be expressed either in terms of exponential functions or in terms of hyperbolic sine and cosine functions. In the lower region where $y < h$, the x and z components are equal to zero on the ground plane and thus are proportional to the hyperbolic sine function. To simplify later calculations, these components can be expressed as

$$\tilde{E}_{x1}(\alpha, y, \beta) = c_x \frac{\sinh(u_1 y)}{\sinh(u_1 h)} \tag{11-42}$$

and

$$\tilde{E}_{z1}(\alpha, y, \beta) = c_z \frac{\sinh(u_1 y)}{\sinh(u_1 h)}, \tag{11-43}$$

where the c coefficients are functions of α and β. Because the divergence of the electric field is zero, the y component can be evaluated as

$$\tilde{E}_{y1}(\alpha, y, \beta) = -\frac{j\alpha c_x + j\beta c_z}{u_1} \frac{\cosh(u_1 y)}{\sinh(u_1 h)}. \tag{11-44}$$

In the region above the interface ($y > h$), the transforms of the x and z components of the electric field can be written

$$\tilde{E}_{x0} = c_x e^{-u_0(y-h)} \tag{11-45}$$

and

$$\tilde{E}_{z0} = c_z e^{-u_0(y-h)}, \tag{11-46}$$

where we have used the continuity of the tangential electric field across the interface. Again, the fact that the divergence of the electric field is zero can be used to show that

$$\tilde{E}_{y0} = \frac{j\alpha c_x + j\beta c_z}{u_0} e^{-u_0(y-h)}. \tag{11-47}$$

From the Maxwell curl-**E** equation, the transforms of the tangential magnetic field components can be shown to satisfy

$$j\omega\mu\tilde{H}_x = -\frac{\partial}{\partial y}\tilde{E}_z + j\beta\tilde{E}_y \tag{11-48}$$

and

$$j\omega\mu\tilde{H}_z = -j\alpha\tilde{E}_y + \frac{\partial}{\partial y}\tilde{E}_x. \tag{11-49}$$

The components below the interface can then be evaluated as

$$j\omega\mu\tilde{H}_{x1} = -\frac{\alpha\beta c_x + (u_1^2 - \beta^2)c_z}{u_1}\frac{\cosh(u_1 y)}{\sinh(u_1 h)} \tag{11-50}$$

and

$$j\omega\mu\tilde{H}_{z1} = \frac{(u_1^2 - \alpha^2)c_x - \alpha\beta c_z}{u_1}\frac{\cosh(u_1 y)}{\sinh(u_1 h)}, \tag{11-51}$$

and the corresponding components for the magnetic field above the interface are

$$j\omega\mu\tilde{H}_{x0} = \frac{-\alpha\beta c_x(u_0^2 - \beta^2)c_z}{u_0}e^{-u_0(y-h)} \tag{11-52}$$

and

$$j\omega\mu\tilde{H}_{z0} = -\frac{(u_0^2 - \alpha^2)c_x - \alpha\beta c_z}{u_0}e^{-u_0(y-h)}. \tag{11-53}$$

At the interface, as y approaches h, these components simplify to

$$j\omega\mu\tilde{H}_{x1}(\alpha, h, \beta) = -\frac{-\alpha\beta c_x + (u_1^2 - \beta^2)c_z}{u_1}\coth(u_1 h), \tag{11-54}$$

$$j\omega\mu\tilde{H}_{z1}(\alpha, h, \beta) = \frac{(u_1^2 - \alpha^2)c_x - \alpha\beta c_z}{u_1}\coth(u_1 h), \tag{11-55}$$

$$j\omega\mu\tilde{H}_{x0}(\alpha, h, \beta) = \frac{-\alpha\beta c_x + (u_0^2 - \beta^2)c_z}{u_0}, \tag{11-56}$$

and

$$j\omega\mu\tilde{H}_{z0}(\alpha, h, \beta) = -\frac{(u_0^2 - \alpha^2)c_x - \alpha\beta c_z}{u_0}. \tag{11-57}$$

The tangential H-field is discontinuous at the $y = h$ interface in accordance with the equations

$$\tilde{H}_{z0} - \tilde{H}_{z1} = \tilde{J}_x \tag{11-58}$$

and

$$-\tilde{H}_{x0} + \tilde{H}_{x1} = \tilde{J}_z, \tag{11-59}$$

which lead to

$$\frac{(u_0^2 - \alpha^2)c_x - \alpha\beta c_z}{u_0} + \frac{(u_1^2 - \alpha^2)c_x - \alpha\beta c_z}{u_1}\coth(u_1 h)$$
$$= -j\omega\mu\tilde{J}_x \tag{11-60}$$

and

$$\frac{-\alpha\beta c_x + (u_0^2 - \beta^2)c_z}{u_0} + \frac{(u_1^2 - \alpha^2)c_x - \alpha\beta c_z}{u_1}\coth(u_1 h)$$

$$= -j\omega\mu\tilde{J}_z. \qquad (11\text{-}61)$$

These are two linear algebraic equations in the two unknown parameters c_x and c_z, and making use of the relations

$$c_x = \tilde{E}_x(\alpha, h, \beta) \qquad (11\text{-}62)$$

$$c_y = \tilde{E}_y(\alpha, h, \beta), \qquad (11\text{-}63)$$

the results are

$$-j\omega\tilde{E}_x(\alpha, h, \beta)$$

$$= \left(\alpha^2\tilde{J}_x + \alpha\beta\tilde{J}_z\right)\frac{u_1 + u_0\coth u_1 h}{\epsilon_0(u_1 + u_0\epsilon_r\coth u_1 h)(u_0 + u_1\coth u_1 h)}$$

$$+ \omega^2\tilde{J}_x\frac{\mu}{u_0 + u_1\coth u_1 h} \qquad (11\text{-}64)$$

and a similar equation for $\tilde{E}_z$. With the notation

$$\tilde{G}_e = \frac{u_1 + u_0\coth u_1 h}{\epsilon_0(u_1 + u_0\epsilon_r\coth u_1 h)(u_0 + u_1\coth u_1 h)} \qquad (11\text{-}65)$$

and

$$\tilde{G}_m = \frac{\mu}{u_0 + u_1\coth u_1 h}, \qquad (11\text{-}66)$$

the equations can be written

$$-j\omega\tilde{E}_x(\alpha, h, \beta) = \left(\alpha^2\tilde{J}_x + \alpha\beta\tilde{J}_z\right)\tilde{G}_e(\alpha, \beta) + \omega^2\tilde{J}_x\tilde{G}_m(\alpha, \beta) \qquad (11\text{-}67)$$

and

$$-j\omega\tilde{E}_z(\alpha, h, \beta) = \left(\alpha\beta\tilde{J}_x + \beta^2\tilde{J}_z\right)\tilde{G}_e(\alpha, \beta) + \omega^2\tilde{J}_z\tilde{G}_m(\alpha, \beta). \tag{11-68}$$

In taking the inverse Fourier transform, the factors α and β introduce derivatives, which can be handled in several ways. One way is associate the derivatives with the entire term, which leads to

$$\begin{aligned} -j\omega E_x(x, h, z) = &-\frac{\partial^2}{\partial x^2}\iint G_e(r)J_x(x', z')\,dx'\,dz' \\ &-\frac{\partial^2}{\partial x\,\partial z}\iint G_e(r)J_z(x', z')\,dx'\,dz' \\ &+\omega^2\iint G_m(r)J_x(x', z') \end{aligned} \tag{11-69}$$

and a similar equation for E_z. A better way numerically probably is to associate at least one of the derivatives with the currents J_x and J_z. When the moment method is used, this requires the use of continuous basis functions. If both derivatives are associated with the currents, the basis functions must have continuous derivatives. With one derivative associated with the currents,

$$\begin{aligned} -j\omega E_x(x, h, z) = &-\frac{\partial}{\partial x}\iint G_e(r)\frac{\partial J_x}{\partial x}(x', z')\,dx'\,dz' \\ &-\frac{\partial}{\partial x}\iint G_e(r)\frac{\partial J_z}{\partial z}(x', z')\,dx'\,dz' \\ &+\omega^2\iint G_m(r)J_x(x', z')\,dx'\,dz'. \end{aligned} \tag{11-70}$$

This equation and the corresponding equation for E_z can then be written in vector form as

$$\begin{aligned} -j\omega\mathbf{E}(x, h, z) = &-\nabla_t\iint G_e(r)\nabla_t'\cdot\mathbf{J}(x', z')\,dx'\,dz' \\ &+\omega^2\iint G_m(r)\mathbf{J}(x', z')\,dx'\,dz'. \end{aligned} \tag{11-71}$$

These equations frequently are written in a different form. The form selected here is consistent with

$$\mathbf{E} = -\nabla\phi - j\omega\mathbf{A}, \tag{11-72}$$

which in the x-direction is

$$E_x = -\frac{\partial\phi}{\partial x} - j\omega A_x. \tag{11-73}$$

The Fourier transforms of these components are

$$\tilde{E}_x = -j\alpha\tilde{\phi} - j\omega\tilde{A}_x, \tag{11-74}$$

where the transforms of the potentials are

$$\tilde{\phi} = \tilde{G}_e\tilde{\rho}_s \tag{11-75}$$

and

$$\tilde{\mathbf{A}} = \tilde{G}_m\tilde{J}, \tag{11-76}$$

which leads to

$$\tilde{E}_x = -j\alpha\tilde{G}_e\tilde{\rho}_s - j\omega\tilde{G}_m\tilde{J}_x \tag{11-77}$$

because

$$j\omega\rho_s = -\frac{\partial J_x}{\partial x} - \frac{\partial J_z}{\partial z}, \tag{11-78}$$

which in terms of the transforms is

$$\tilde{\rho}_s = \frac{-j\alpha\tilde{J}_x - j\beta\tilde{J}_z}{j\omega}. \tag{11-79}$$

Substitution of this expression gives

$$\tilde{E}_x = \tilde{G}_e \frac{\alpha^2 \tilde{J}_x + \alpha\beta \tilde{J}_z}{j\omega} - j\omega \tilde{G}_m \tilde{J}_x. \tag{11-80}$$

Comparison with

$$\begin{aligned}\tilde{E}_x(\alpha, h, \beta) = &-\frac{\alpha^2 \tilde{J}_x + \alpha\beta \tilde{J}_z}{j\omega} \\ &\times \frac{u_1 + u_0 \coth u_1 h}{\epsilon_0 (u_1 + u_0 \epsilon_r \coth u_1 h)(u_0 + u_1 \coth u_1 h)} \\ &+ j\omega \tilde{J}_x \frac{\mu}{u_0 + u_1 \coth u_1 h}\end{aligned} \tag{11-81}$$

shows that

$$\tilde{G}_e = -\frac{1}{j\omega} \frac{u_1 + u_0 \coth u_1 h}{\epsilon_0 (u_1 + u_0 \epsilon_r \coth u_1 h)(u_0 + u_1 \coth u_1 h)} \tag{11-82}$$

and

$$\tilde{G}_m = -\frac{\mu}{u_0 + u_1 \coth u_1 h}. \tag{11-83}$$

CHAPTER TWELVE

Mode Matching

12.1. APPROXIMATING A FUNCTION BY A SERIES

A fundamental problem in matching is approximation of a function that is approximated by a finite series, which in one dimension can be written

$$g(x) \simeq \sum_{j=0}^{N} g_j B_j(x). \tag{12-1}$$

Frequently, the finite series is obtained by truncation of an infinite series that should equal the function. The problem is to select the coefficients g_j so as make the approximation in some sense a good approximation. We have already encountered this problem as part of the moment method, but it is encountered in a wider context. A fairly general approach is to select a set of $N + 1$ weighting functions $w_i(x)$, multiply both sides, and perform an integration over the domain of x, resulting in

$$\int w_i(x) g(x)\, dx \simeq \sum_{j=0}^{N} g_j \int w_i(x) B_j(x)\, dx. \tag{12-2}$$

As discussed in Chapter 7, this method is frequently referred to as the *method of weighted residuals*, since if the residual is defined as

$$R(x) = g(x) - \sum_{j=0}^{N} g_i B_j(x), \tag{12-3}$$

the weighted residual is equal to zero if

$$\int w_i(x) \left[g(x) - \sum_{j=0}^{N} g_j B_j(x) \right] dx = 0, \tag{12-4}$$

which is equivalent to the equation derived earlier. A specific set of weighting functions sometimes used is a set of delta functions

$$w_i(x) = \delta(x - x_i), \tag{12-5}$$

which leads to the point-matching result

$$g(x_i) \simeq \sum_{j=0}^{N} g_j B_j(x_i). \tag{12-6}$$

The Galerkin selection of weighting function was another approach discussed when the moment method was covered. A similar approach is the *least square error* (LSE), where the coefficients are so selected as to minimize the integrated squared error

$$\text{ISE} = \int \left[g(x) - \sum_{j=0}^{N} g_j B_j(x) \right]^2 dx. \tag{12-7}$$

The minimum value of this expression is defined by the set of equations

$$\frac{\partial}{\partial q_i} \int \left[g(x) - \sum_{j=0}^{N} g_j B_j(x) \right]^2 dx = 0, \qquad (12\text{-}8)$$

which leads to

$$\int B_i(x) g(x)\, dx = \sum_{j=0}^{N} g_j \int B_i(x) B_j(x)\, dx. \qquad (12\text{-}9)$$

Comparison with the equations above shows that this approach is equivalent to selecting the weighting functions as

$$w_i(x) = B_i(x). \qquad (12\text{-}10)$$

If the functions are complex, a better approach is to define the integrated squared error by the integral of the square of the magnitude of the error

$$\text{ISE} = \int \left[g(x) - \sum_{j=0}^{N} g_j B_j(x) \right] \left[g(x) - \sum_{j=0}^{N} g_j B_j(x) \right]^* dx, \qquad (12\text{-}11)$$

from which the resulting equations can be shown to be

$$\int B_i^*(x) g(x)\, dx = \sum_{j=0}^{N} g_j \int B_i^*(x) B_j(x)\, dx. \qquad (12\text{-}12)$$

12.2. INDUCTIVE IRIS IN A WAVEGUIDE (REVISITED)

We will now apply the method of mode matching to the same inductive iris in a waveguide problem considered before. The incident field is given by

$$E_y^i(x, z) = \sin \frac{\pi x}{a} e^{-\gamma_1 z}, \tag{12-13}$$

where γ_1 is

$$\gamma_1 = \sqrt{\left(\frac{\pi}{a}\right)^2 - k_0^2}. \tag{12-14}$$

The corresponding magnetic field is described by

$$H_x^i(x, z) = -\frac{\gamma_1}{j\omega\mu_0} \sin \frac{\pi x}{a} e^{-\gamma_1 z}. \tag{12-15}$$

The reflected (scattered) wave is described by

$$E_y^r(x, z) = \sum_{n=1}^{\infty} e_n \sin \frac{n\pi x}{a} e^{\gamma_n z} \tag{12-16}$$

and

$$H_y^r(x, z) = \sum_{n=1}^{\infty} \frac{\gamma_n}{j\omega\mu_0} e_n \sin \frac{n\pi x}{a} e^{\gamma_n z}. \tag{12-17}$$

Because the total electric field is zero on the iris

$$E_y^i(x,0) + E_y^r(x,0) = 0 \qquad \text{for} \quad 0 < x < c, \tag{12-18}$$

which can be written

$$\sum_{n=1}^{\infty} e_n \sin \frac{n\pi x}{a} = -\sin \frac{\pi x}{a} \qquad \text{for} \quad 0 < x < c, \tag{12-19}$$

and because the tangential magnetic field is continuous where there is no metal

$$\sum_{n=1}^{\infty} \frac{\gamma_n}{j\omega\mu_0} e_n \sin \frac{n\pi x}{a} = 0 \qquad \text{for} \quad c < x < a. \tag{12-20}$$

If we terminate the summation at N, we get the approximations

$$\sum_{n=1}^{N} e_n \sin \frac{n\pi x}{a} \simeq -\sin \frac{\pi x}{a} \qquad \text{for} \quad 0 < x < c \tag{12-21}$$

and

$$\sum_{n=1}^{N} \frac{\gamma_n}{j\omega\mu_0} e_n \sin \frac{n\pi x}{a} \simeq 0 \qquad \text{for} \quad c < x < a. \tag{12-22}$$

This is a special case of the problem discussed in Section 12.1. We could, for example, divide the interval $(0, a)$ into N subintervals and use pulse functions for weighting, with N_1 pulse functions for the interval $(0, c)$ and with $N_2 = N - N_1$ for the interval (c, a).

12.3. OTHER EXAMPLES OF WAVEGUIDE MODE MATCHING

The concept of mode matching has been applied to a variety of waveguide problems. Any circuit that consists of a number of connected sections, each of which is simple enough to have a solution involving known modes, usually can be analyzed by enforcing continuity of tangential field components at the interfaces between the sections. A change in width of a rectangular waveguide carrying a TE_{01} mode, as illustrated in Fig. 12-1, offers a simple example. The E and H fields on each side of the interface where

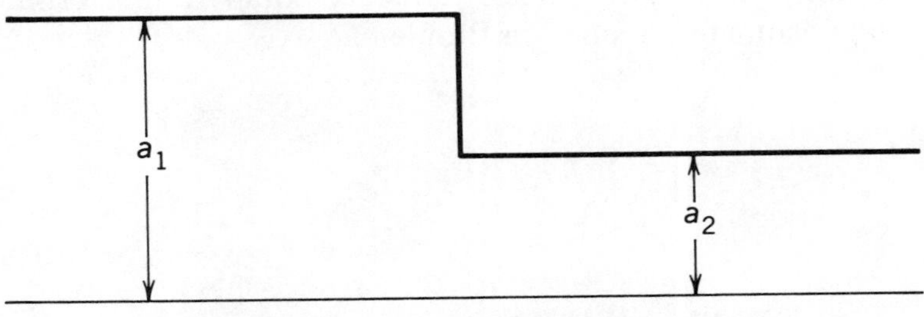

FIGURE 12-1. Change in width of a waveguide.

the width changes easily can be expanded in infinite series involving well-known modal functions. Equating the tangential field components at the interface and terminating the summations to a finite number of terms yield several approximate equalities. Use of weighting functions, as discussed in Section 12.1, yields a set of linear algebraic equations.

PROBLEMS

1. For the waveguide width change illustrated in Fig. 12-1, write the appropriate modal expansions and indicate how a set of algebraic equations for the coefficients can be derived.
2. For the situation in Fig. 12-1, assume an incident wave from the left. Using just the dominant modes and unity weighting, calculate the reflection coefficient.

COMPUTER PROJECT 12-1

For the situation shown in Fig. 12-1, with

$$a_1 = 5 \text{ cm}$$
$$a_2 = 4 \text{ cm}$$

assume an incident wave from the left which is a TE_{10} mode at a frequency of 5 GHz. Write a computer program to determine the reflection coefficient. Use eight modes on each side, and match at the discontinuity using weighting functions which are sinusoids that correspond to the modes on the left side. Compare the result with the crude estimate obtained in Problem 2.

Concluding Comments

In this book we have described several basic methods of numerical analysis for solving electromagnetic problems. As can be seen from other books and the professional literature, many other methods have been derived. Two of these are the transmission line matrix (TLM) method—a time-domain method similar to the finite-difference time-domain method and essentially a numerical version of Huygen's principle—and the boundary-element method—a variation of the finite-element method in which the boundary rather than the interior region is divided into elements. Additionally, many other new methods, most of which are combinations and variations on older methods, still are being derived. While much of this activity represents real progress, the variety of methods certainly is likely to be confusing. One question frequently arises: "What is the best method for my particular problem?" The literature contains little in the way of real comparisons of methods, and what does appear is usually contradictory. The most important criteria in comparing methods are the time and effort to derive the algorithm and write the program, computer memory required, and computer run time required to achieve desired accuracy. This last criterion should receive more attention in the future. It seems to me that if there were really a superior method and everyone were rational, this method would be the only one used and the others would die away. Clearly this has not happened and hence a logical conclusion would seem to be that different problems are best solved by different methods.

Although computational electromagnetics has a fairly long history, it still is a very active area of research and undoubtedly will continue to be. Most papers on numerical methods for electromagnetics are found in the three *IEEE Transactions on Antennas and Propagation*, *Microwave Theory and Techniques*, and *Magnetics*, and the interested reader is encouraged to read these journals and others, such as *Radio Science*. Improvements in computer hardware and software have influenced the development of the subject and expected improvements, especially those involving parallel computation, promise an exciting future.

Index